T0130979

Broadband Wireless Communications Business

Broadband Wireless Communications Business

An Introduction to the Costs and Benefits of New Technologies

Riaz Esmailzadeh
IPMobile Inc., Japan

John Wiley & Sons, Ltd

Telephone (+44) 1243 779777

Email (for orders and customer service enquiries): cs-books@wiley.co.uk
Visit our Home Page on www.wiley.com

Other Wiley Editorial Offices

John Wiley & Sons Inc., 111 River Street, Hoboken, NJ 07030, USA

Jossey-Bass, 989 Market Street, San Francisco, CA 94103-1741, USA

Wiley-VCH Verlag GmbH, Boschstr. 12, D-69469 Weinheim, Germany

John Wiley & Sons Australia Ltd, 42 McDougall Street, Milton, Queensland 4064, Australia

John Wiley & Sons (Asia) Pte Ltd, 2 Clementi Loop #02-01, Jin Xing Distripark, Singapore 129809

John Wiley & Sons Canada Ltd, 22 Worcester Road, Etobicoke, Ontario, Canada M9W 1L1

Wiley also publishes its books in a variety of electronic formats. Some content that appears in print may not be available in electronic books.

Library of Congress Cataloging-in-Publication Data

Esmailzadeh, Riaz.
Broadband wireless communications business : an introduction to the costs and benefits of new technologies / Riaz Esmailzadeh.
p. cm.
Includes bibliographical references and index.
ISBN-13: 978-0-470-01311-3
ISBN-10: 0-470-01311-7
1. Information technology – Economic aspects. 2. Telecommunication – Economic aspects. 3. Broadband communication systems. 4. Wireless communication systems. I. Title.

HC79.155.E84 2006
384.3 – dc22

2005056960

British Library Cataloguing in Publication Data

A catalogue record for this book is available from the British Library

ISBN-13 978-0-470-01311-3 (HB)
ISBN-10 0-470-01311-7 (HB)

Typeset in 10/12pt Times by Laserwords Private Limited, Chennai, India.
Printed and bound in Great Britain by Antony Rowe Ltd, Chippenham, Wiltshire.
This book is printed on acid-free paper responsibly manufactured from sustainable forestry in which at least two trees are planted for each one used for paper production.

To Izumi, Amin Asad, and Kian

Contents

List of Figures

List of Tables

Preface

The past two decades have witnessed the introduction and unprecedented growth of cellular mobile telephony and wireless communications. Numerous wireless access technologies have been introduced in the mobile communications market. While some have flourished and formed the basis of successful manufacturing and network operator businesses, many have lived only for a short time and disappeared. With the saturation of the market in most developed and many developing countries, attention has been shifted to the provision of broadband wireless access to the internet, as well as multimedia services that require high transmission rates. Again, several technologies are candidates for this new phase of market development. What the characteristics of these technologies are and how they relate to what it takes to become successful in this market are the topics this book addresses.

As with most other businesses, wireless and mobile communications systems must operate within limited available resources. From a technical point of view, these include frequency bandwidth, transmission power and battery capacity, processing power, size of devices, and so on. Among these, frequency bandwidth directly relates to the maximum possible system transmission rate and user capacity; and maximum transmission power determines the extent of the area within which services may be provided. The expected diversity of future services also generates a new set of requirements on forms and functionalities of devices. How broadband wireless technology and devices best utilise these resources are of major consequence to their success in the market place. In fact, one major research and development focus for radio communications technology, for over a century, has been on increasing the utilisation efficiency of these resources.

The most relevant technologies can be classified into several groups. In this book, we address the following: (1) access technologies, which above all define how wireless communications devices share the frequency bandwidth resource; (2) associated enhancing technologies, such as specific processing technologies within the devices, and antenna technologies which help realise a more efficient usage of transmission power while minimising interference to other devices; and (3) network topologies which enable the provision of wireless communications services over a desired coverage area. The access technologies can be combined with one or more of the enhancing technologies and networking topologies to increase overall system efficiency. However, not all of these technologies can be combined. The specific character of an access technology may mean some enhancing technologies cannot be utilised. The ability of an access technology to be combined with one or more associated technologies can lead to advantages in system performance, which may spell success or failure in the market place.

This book first gives a summarised background of the wireless and mobile communications technologies in Chapter 1. It reviews the present major access technologies, provides a short history of how they have come about, and gives examples of where they are most utilised. The book then gives a background of wireless communications and the challenges involved in serving a large number of users in a wireless environment, and their dependance on the characteristics of communications channels, the frequency band of operation, and the losses incurred when transmitting over the air. This background is used to explain why major present technologies, such as PDC, GSM, CDMA, WCDMA, TD-CDMA, WLAN, WiMAX, and so on, have appeared. Then, in Chapter 2, the focus is on a subset of these technologies and their technical characteristics, how they relate to the cost of network layout, and to the operation. The issues of costs associated with each technology are discussed. The characteristics of each system in facilitating multi-user shared utilisation of resources are discussed and compared, leading to a discussion of the capacities of each technology and the number of users that can be served by each network.

In Chapter 3, techniques that can enhance the performance of the major wireless communications technologies are described. First, the technologies for interference mitigation are reviewed with considerations of operating cost and performance. Next, antenna technologies that affect the total system performance in terms of power efficiency, higher transmission rates, and larger user capacities are discussed. On the basis of these, conclusions are made on how cost savings in network operation can be realised.

Chapter 4 deals with wireless network topologies. Here, centralised base stations, as used, for example, in public mobile communications, versus distributed access points, as used, for example, in WLAN systems, are discussed. This is followed by a discussion of the new network topologies, such as ad hoc networking and how these create new business models. The different segments of the market, public mobile systems, hot spot coverage, personal area network, and so on, are then addressed. The technologies needed to address these configurations are defined.

In the next part of this book, Chapters 5–7, issues of costs associated with building and operating a broadband wireless communications network are discussed. The cost of spectrum (Chapter 5), the cost of devices such as base station and end-user equipment (Chapter 6), and the costs of design and operating a cellular communications network (Chapter 7) are dealt with.

Chapter 8 discusses the services that can be provided over the broadband wireless communications network; how an operator can gain revenues from communication services; what services are expected to be carried over by these networks at present and in future; and possibilities of how revenues might flow to all players in the value chain.

In the final chapter, several business scenarios, or business plans, are developed for fictitious operators in order to illustrate how new technologies can be used to provide broadband wireless communications services in two markets, Japan and China.

The book focuses primarily on the major present technologies, which are considered for broadband wireless access: namely, WCDMA, WLAN, TD-CDMA, and WiMAX. Throughout the book, these technologies are compared with reference to the cost of manufacturing, implementation, network design and operation, handsets, and data rates in order to show how each component adds value to the whole.

A Further Reading list has been included at the end of each chapter for readers who need more information on technical or business aspects of the topics discussed in the chapter.

The reader will find this book a valuable reference for:

- Choice of technology: business decisions on which technology to use.
- How to combine several technologies to reach a target market.
- How to differentiate from competitors.
- How to understand and evaluate a technology and how to take advantage of future possible enhancements.
- Which technologies to look for and how to position an operation.
- The value chain and which part of the market may be of value to a particular business.

We are in a period of diverging technologies. After years of standardised network operation, we now have several worldwide standards, as well as related technologies that have spread without becoming an international standard. Increasingly, higher bit rates have allowed wireless communications to become a direct competitor of fixed-line services based on technologies such as digital subscriber link (DSL). There are signs of convergence: large numbers of diverse technologies cannot survive in an integrated high capital expenditure industry such as wireless communications. Economies of scale, international roaming, and device standards work better if only one or maybe a very few technologies survive. How all these divergent technologies merge, and which will survive, are topics of great interest to the industry. A clear explanation of the technologies, associated with cost/performance analyses and business implications, is valuable to decision makers within the telecommunications industry, who may not necessarily have a technical background. This is an important reason for writing this book. Another reason has been the belief that it is important for the engineer to evaluate each technology, not only on its technical merits, but also on the basis of the business aspects and implicit return on investment. It is hoped this book will prove valuable for these managers and engineers, as well as for their business strategists.

Acknowledgements

The idea for this book emerged through my involvement with the process of technology evaluation for a new mobile operator in Japan. At the time, I was fresh from finishing an Executive MBA, and felt the serious need for a book that defines modern communications technology in terms of its business implications.

To turn the idea into a reality, I have been supported and assisted by many people. Through the years, I have become indebted to many educators and professors. First, I would like to acknowledge Professor Masao Nakagawa of Keio University for his support and encouragement. Then, there is Professor Derek Abell, formerly of IMD, and now of ESMT, who has been a great inspiration, guiding me on the topic of technology management. Also, I have been inspired by discussions with colleagues at IPMobile Inc., including Messrs Hiroshi Takeuchi, Koichi Maruyama, Kenji Nobukuni, Kazunari Takeuchi, Isao Yoshikawa, and Tsutomu Kajiya, all of whom have supported me through their valuable comments. I would also to like to acknowledge Dr Masanobu Fujioka, Mr Tetsuo Takasago, Dr Hiroshi Mano, Mr Steve Roberts among many for many hours of discussion. I also appreciate the permission granted by Professor Fumiyuki Adachi, IEEE, DSL Forum, John Wiley and Sons, KDDI, OBSAI, WWRF, ITMedia, and UMTS Forum to reproduce their copyrighted material here. I owe my gratitude to John Schwerin for patiently reviewing and proofreading the English text.

I am also thankful to the editors at John Wiley and Sons, Ltd. including Mark Hammond, Sarah Hinton and Olivia Underhill, and to Alistair Smith from Sunrise Setting, Ltd. for their help throughout this project.

Most of all, I am grateful to my wife and two sons, who have patiently endured my preoccupation with this work while at home. To them, I dedicate this book.

1

Background

Broadband wireless communications technologies promise the freedom of being constantly connected to the internet at high speeds without the limitation of connection cables. There is certainly a need for such a service and, therefore, there is a market. The size of the market depends on the cost of the technologies that can realise the service. As in any business, this cost is passed on to the end-user in the form of user equipment (e.g. mobile phone) price and usage tariffs. The lower the end-user cost is, the greater the market size becomes.

But how have we arrived at this point? Why have these technologies been developed? Why are there so many of them? (See Box: Broadband wireless technologies.) And what factors determine which technologies will ultimately emerge to dominate the market?

Some broadband wireless technologies

- WCDMA, standardised by 3GPP
- CDMA-2000, standardised by 3GPP2
- TD-CDMA, standardised by 3GPP
- Wireless LAN IEEE 802.11, standardised by IEEE
- WiMAX IEEE 802.16, currently being standardised by IEEE
- TD-SCDMA(MC), proprietary technology by Navini
- i-Burst, proprietary technology by Kyocera, Arraycomm
- Flash-OFDM, proprietary technology by Flarion

This book discusses the two parts of the broadband system as shown in Figure 1.1. One is the interface between the end-user device and a central station. In fixed-line systems,

Broadband Wireless Communications Business: An Introduction to the Costs and Benefits of New Technologies Riaz Esmailzadeh

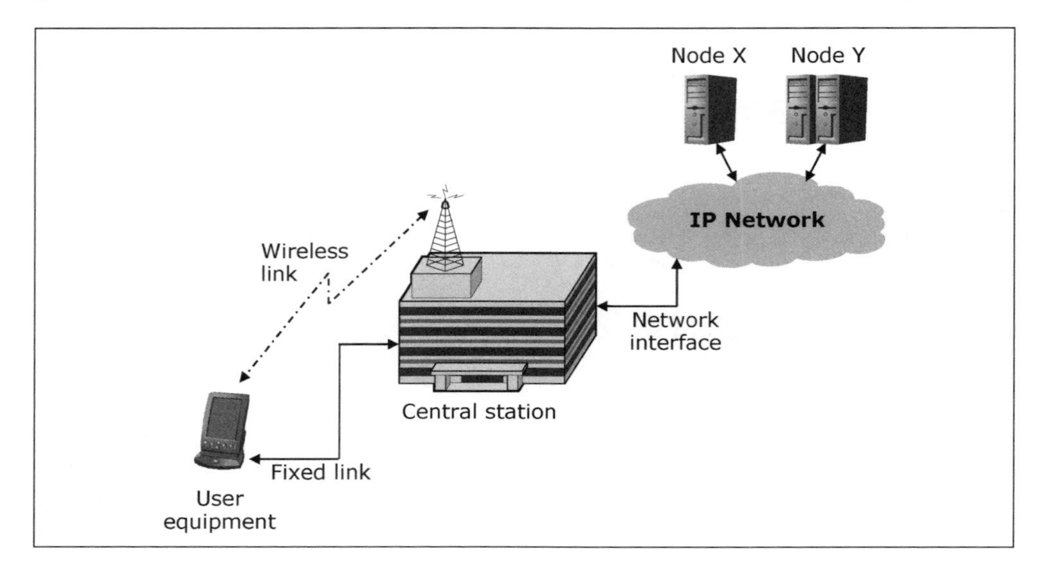

Figure 1.1 Interfaces for access to the internet

the central station may be a telephone exchange or a central office, or perhaps a TV cable operations centre that also provides internet connections. The connection may be an optical fibre or a twisted pair of copper wires. In wireless systems, the central station is, for example, a WCDMA base station. The connection is wireless, and the interface is referred to as *air interface*. The central station connects to the internet, usually through several intermediary devices or nodes. This is referred to as *network interface*. Most of the technologies discussed in this book concern the air interface. However, we will also discuss the network interface as this too significantly affects the systems' costs.

Before discussing the technologies and their relative merits and costs, let us go through a brief background on how broadband wireless communications have come about. In this chapter, we discuss the background of the two apparent parts of broadband wireless technology: (1) broadband access to the internet as developed for fixed communications; and (2) wireless communications as developed for cellular applications. The combination of these two fields has come about to address two particular needs: one is tetherless access to the internet and the other is access to the internet where the fixed option is impossible or comparatively uneconomical. This includes technologies that provide access to high-transmission-rate services, which are of particular benefit to a mobile end-user. In this chapter, the background will be discussed as follows:

- Fixed-line data communications development
- Broadband access market and its growth
- Mobile communications market, its growth, and evolution

- Wireless data communications
- Present status of wireless market.

From these, we draw conclusions on trends for future developments.

1.1 Fixed-line Data Communications

Computers connect to the internet in several ways. A large group of computers may use a local area network (LAN) to communicate with each other, to communicate with peripheral devices such as databases and printers, and to communicate with the internet. A home personal computer (PC) may dial up the internet using a normal phone line or may use a fixed cable that is connected to the net (often the same as that of a cable TV system), and so on. Fast connections to the internet from home were not readily available until very recently (see Box: Modem rates evolution). Fast connections have existed for larger, office-like environments for a long time. However, until very recently, connecting from the home was usually possible using these modems with a maximum rate of a few tens of kilobits per second (kbps). An alternative, Integrated Services Digital Network (ISDN) (see Box: Integrated services digital network) was standardised by the International Telecommunications Union (ITU) in the 1980s to provide up to 144 kbps transmission rates over telephone

Modem rates evolution

1981	Hayes Smartmodem	300 bps
Late 1980s		2.4 kbps
1991	Superfax 14400	14.4 kbps
Late 1990s		56 kbps

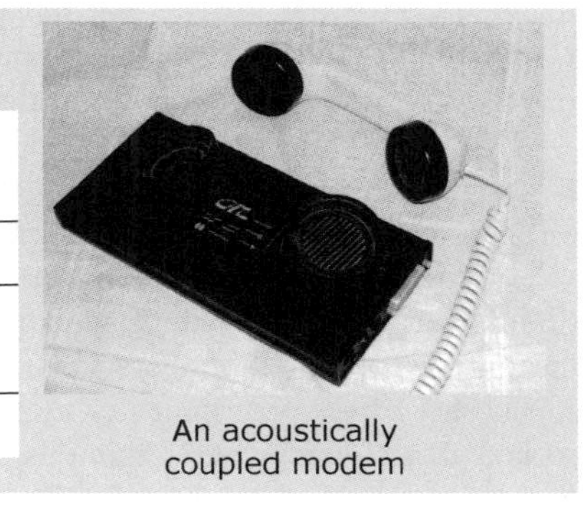

An acoustically coupled modem

Integrated services digital network

Original recommendations of ISDN were in CCITT Recommendation I.120 (1984). ISDN services did not become widespread owing to some differences in regional standards, delays in establishing a market and then reaching that market, which was soon better served by DSL. For example in Japan, ISDN services started in mid-1990s, flat-rate services started in late 90s (e.g. Flet's ISDN by NTT) and the subscriber numbers were growing. However, the emergence of DSL services almost at the same time (2000) ended that growth.

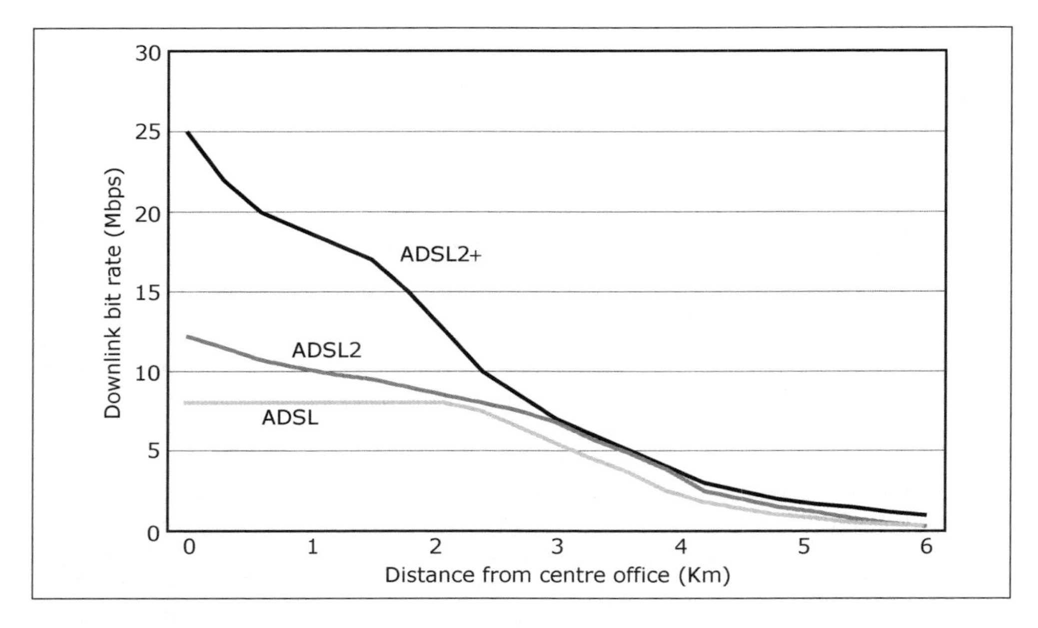

Figure 1.2 DSL rates versus distance. Reproduced by permission of DSL Forum

lines. However, the service was supported by only a few operators around the world. It was not until the late 1990s when the development of digital subscriber links (DSL) made inexpensive fast connections from the home possible.

DSL technology utilises the standard telephone's twisted copper-wire phone connection. Because of losses over these links, the maximum transmission rates are a function of the distance between the telephone exchange and homes. Figure 1.2 shows the possible rates versus distance. As the distance between exchanges themselves is a function of population density, the denser the population, the higher the density of exchanges and therefore the higher the possible average DSL rates. This means that while in countries such as Japan and Korea very high DSL rates of tens of Mbps are being offered, DSL operators in less densely populated countries offer services at 256–512 kbps. Latest DSL systems do offer rates in the order of several Mbps, but still only for users near the exchanges. The transmission rates for users separated from the station by more than 5 kilometers is very low.

Higher transmission rates, even in excess of 100 Mbps, are available with fibre-to-the-home (FTTH) solutions. These are optical fibre connections that directly connect a home to a network, which itself is connected to the internet. This service, however, is almost confined to highly populated urban centres.

1.2 Mobile Communications

'Phenomenal', 'unprecedented', and 'explosive' are but a few of the adjectives that have been used to describe the growth of the mobile communications industry over the past

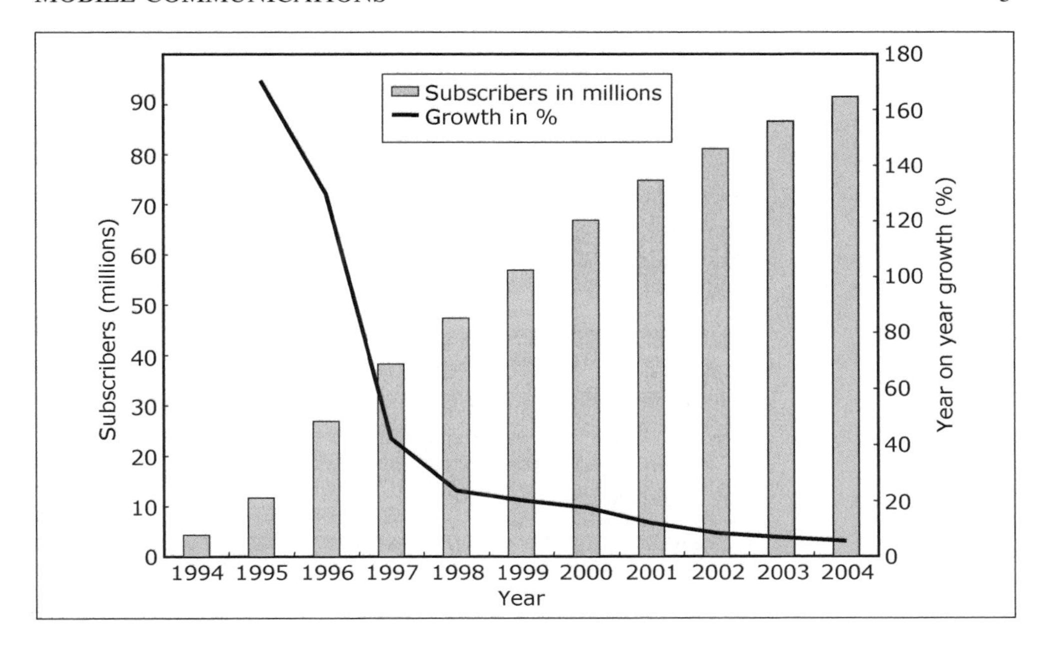

Figure 1.3 Subscriber growth in Japan. Reproduced by permission of Japanese Ministry of Internal Affairs and Communications (MIC)

decade. Modern mobile communications date to the early 1980s, but growth in those early days was extremely slow. Constrained by the high cost and bulky size of mobiles as well as high tariffs, these services were kept out of the reach of most consumers. In the mid-1990s, however, this scenario started to change as smaller and less-expensive handsets started to appear in the market, and operators started to offer affordable service packages. In many developed countries, the growth rate has been so fast that in less than a decade virtually all of their population have become mobile phone subscribers. Figure 1.3 shows how the market has grown in Japan. In less than six years, market penetration grew from less than 10% in 1995 to more than 60% in 2000. Market penetration has been highest among young people – as of April 2004, more than 99% of people in their twenties and 95% of people in their thirties had a mobile phone. Similar growth rates are observed elsewhere in the developed world. Table 1.1 shows the penetration rate for several highly market-saturated countries. Owing to some interesting peculiarities of their mobile market development, there exist countries with more than 100% market penetration. That is, a significant proportion of the population have more than one subscription. The 100%+ penetration rates are partly due to the following: (1) many of the subscribers may be foreign workers; (2) some subscribers have two or more prepaid subscriptions; and (3) coverage for one mobile technology (e.g. GSM) services are not available in the countryside, and there is a need to be connected while on holidays, using older analogue technologies. The rate of growth in developing countries more and more mirrors that of the developed world of earlier years. The compound annual growth rate (CAGR) of several countries is shown in Table 1.2.

Table 1.1 Mobile penetration rate for several countries. *Source*: ITU statistics, March 2005

Country	Market penetration %
Luxembourg	119.4
Taiwan	114.1
Hong Kong	107.9
Italy	101.8
Sweden	98.1

Table 1.2 Compound annual growth rate (CAGR) for subscribers in several countries, over 1998–2003. *Source*: ITU statistics, March 2005

Country	CAGR %
Sweden	16.5
Japan	12.9
Tajikistan	157.6
Jamaica	87.0
Uganda	91.7

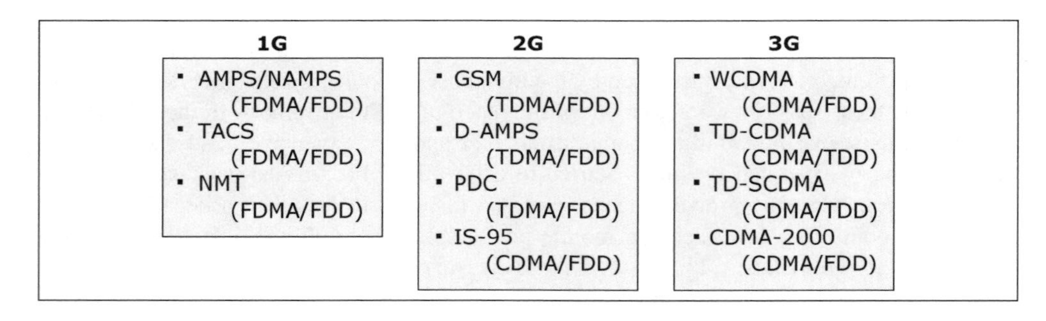

Figure 1.4 Three generations of mobile communications technologies

The technologies for these services are generally considered to have gone through three generations. A typical classification of these technologies is shown in Figure 1.4. While the first-generation devices used analogue technologies and provided primarily voice services, second- and third-generation devices have been digital, and suitable for data communications. Some of these technologies have been jointly developed by a group of companies. They have gone through a thorough evaluation and revision process, and have become worldwide standards. Some of these are now in service or are being commercialised.

Table 1.3 Typical cost figures for new subscribers for 2G and 3G mobiles in Japan

Model	Type	Operator	Manufacturer	Price ($)
FOMA-N901	3G high end	DoCoMo	NEC	110
MOVA-P700i	2G high end	DoCoMo	Panasonic	70
M1000	3G high end with PDA	DoCoMo	Motorola	460
WIN	3G high end	KDDI	Toshiba	130
Au	2G high end	KDDI	Casio	120
Tuka	2G low end	Tuka	Kyocera	40

Meanwhile, some other proprietary technologies have been developed mainly through the efforts of individual companies. First-generation (1G) systems have been almost totally replaced by second-generation (2G) systems, while third-generation (3G) systems are starting to augment, and gradually replace, those of 2G.

All along, advances in microelectronics and the economies of scale have driven down the cost of both mobile handsets and network equipment, allowing operators to offer less-expensive plans to attract more subscribers. The end-user cost of devices has remained very much a function of competition in the marketplace. In many instances, operator subsidy has driven the end-user price to zero. Meanwhile, increasingly abundant and diverse accessory features, such as cameras, large displays, radios, and even televisions, have been packaged in the mobile handset. These new features are reasons why the price of a handset has not fallen even more dramatically. Table 1.3 shows the present cost of different 2G and 3G end-user mobiles in Japan. These are the prices a new subscriber would pay after operator subsidy, which in Japan differs for various models, but on average it is around $400 per handset.

Competition between operators has been the major factor deciding the monthly tariffs, usually expressed as average revenue per user (ARPU). This is the total a subscriber pays to the operator each month for mobile services, including a basic fee, for talk-time charges as calculated per unit of time for voice services, and data communications charges as calculated per unit of data. The ARPU trend in Japan over the past few years is shown in Figure 1.5. Generally, ARPU figures have been falling over the past few years in Japan and most other countries. One important reason has been price competition between operators. Another reason is the fact that most new subscribers are older people, who do not use their mobiles so often. A third factor, as observed from Figure 1.5, is the decline in average voice revenue in comparison to data revenue.

1.3 Wireless Data Communications

Initially, mobile communications systems provided telephony services (paging services also existed, but we do not consider those services in this text). Soon, however, data services emerged. Initially, these were short messaging services (SMS). With the great spread of the internet in the fixed networks, similar services started to appear in the mobile networks. The Japanese i-mode service (see Box: Japanese i-mode service) and Wireless Access Protocol

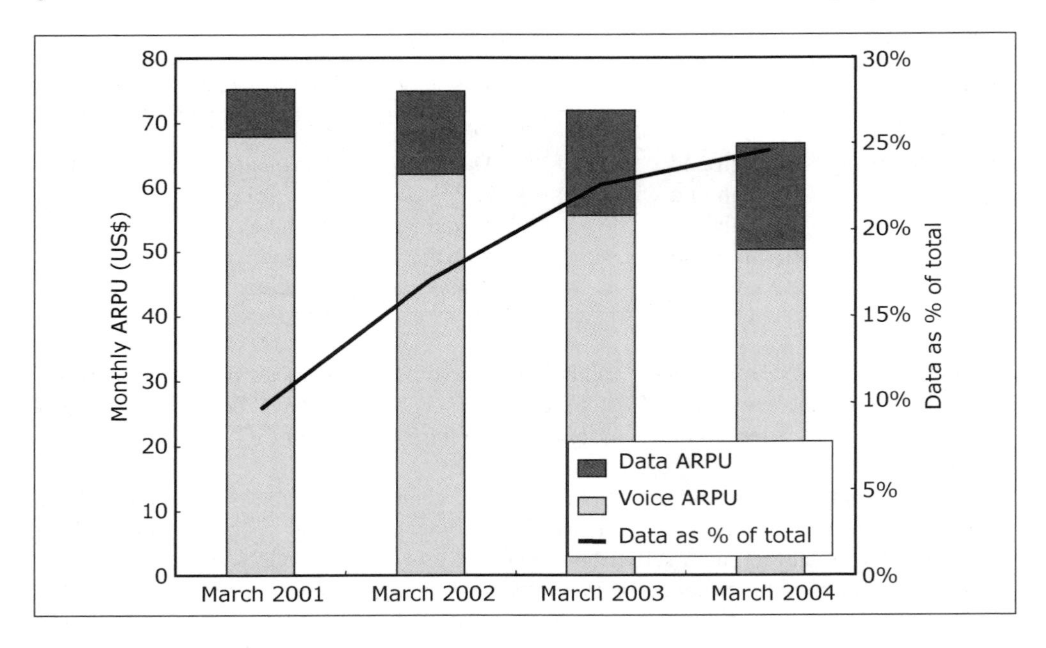

Figure 1.5 ARPU trend in Japan. Reproduced by permission of Japanese Ministry of Internal Affairs and Communications (MIC)

Japanese i-mode service

NTT DoCoMo introduced i-mode in February 1999. Because of low PC penetration at home, and the high cost of dial-up internet connection, i-mode services caught on very quickly. It was also because i-mode provided SMS functionality, which was then not widely available for mobile subscribers.

DoCoMo's use of simplified web-browsing language, a content provider system, and a handset maker alliance ensured the great commercial success of i-mode. Nowadays, i-mode, and similar services by other Japanese operators, enable users to connect to the internet. The percentage of internet-enabled mobiles in Japan is the highest in the world. The success of i-mode in Japan has attracted the attention of operators in other countries. Here are some examples of (mainly) GSM operators who are adopting i-mode:

Mar 2002	E-Plus, Germany
Apr 2002	KPN, The Netherlands
Jun 2002	KG Telecom, Taiwan
Nov 2002	BASE, Belgium
Nov 2003	Wind, Italy
May 2004	Cosmote, Greece

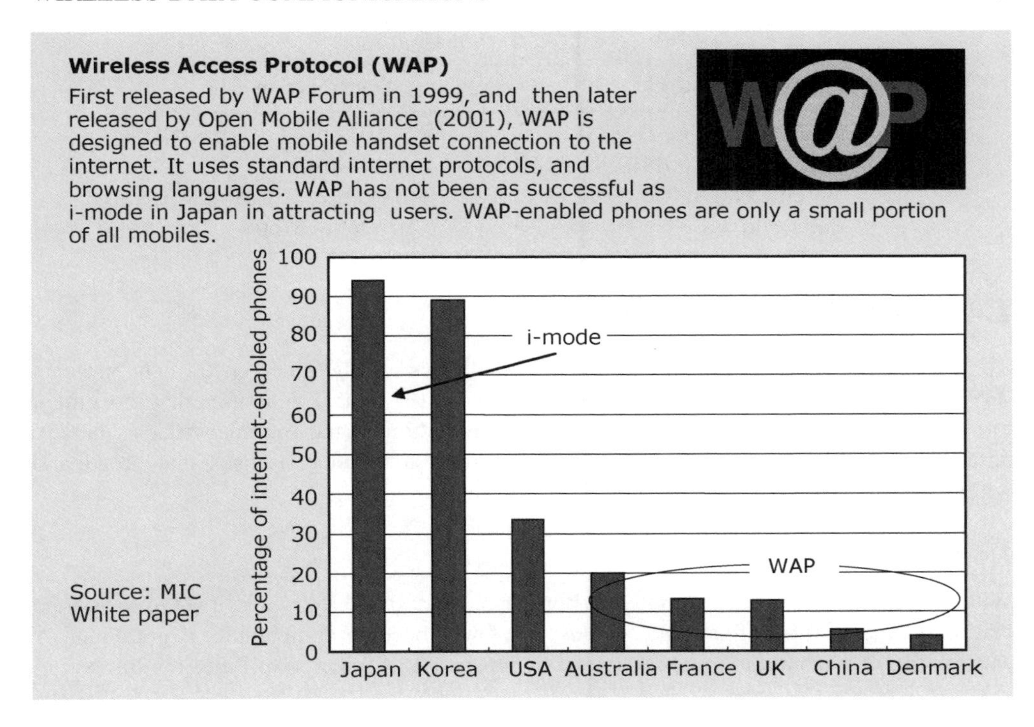

(WAP) (see Box: Wireless Access Protocol (WAP)) are examples of these services. I-mode and other similar services in Japan have been particularly successful because they provided e-mail/short messaging capability, which was not otherwise readily available, as well as provided access to information via their internet browser. Because of its great success in Japan, i-mode has been copied by several European operators, but with mixed results. The general ARPU trend, however, has been the growth of data communications, and the decline of voice communications as shown in Figure 1.5.

Wireless data communications is therefore viewed as the growth component of a mobile operators' business. But this growth can be maintained only if present services are enhanced and new services are introduced. The pace of data ARPU is already slowing down as can be seen form Figure 1.5, perhaps as a result of insufficient new data service offerings. Most new services are likely to include large-sized content, including video and audio, as well as rich-content multimedia. Much higher data rates will be required for these new services and applications, and this has been one of the main drivers of the evolution from 2G to 3G, and a driver for the development of related technologies.

Wireless (mobile) data communications services became available initially with 2G systems. These systems provide transmission rates of up to a few tens of kbps, which, although sufficient for SMS or simple web browsing, cannot support applications with larger file sizes and transmission speed requirements. A major factor in the development of 3G standards has therefore been to provide technologies that can deliver much higher data rates. Table 1.4 lists the possible transmission rates for several present technologies.

Table 1.4 Data transmission rates

Access technology	Transmission rate
2G (GSM, PDC, D-AMPS)	~10 kbps
Enhanced 2G (GPRS, D-AMPS)	~144 kbps
3G (WCDMA, TD-CDMA)	384 kbps ~ 2 Mbps
Enhanced 3G	Up to 14 Mbps

1.4 Broadband Wireless

Considering the saturation of the mobile telephony market, the emergence of mobile data services, and the growth of fixed broadband services, it is logical to assume that movement into broadband wireless services is the natural evolution of the present wireless market. Although expectations of market size vary, there is a broad consensus that broadband wireless is the next 'big thing' for the industry.

1.4.1 Edholm's Law

With the growth of transmission rates on both the fixed and wireless sides, it is fair to assume that sometime in future fixed and wireless rates will become comparable, if not equal. A comparison of transmission rates is shown in Figure 1.6. Attributed to Phil Edholm, Nortel

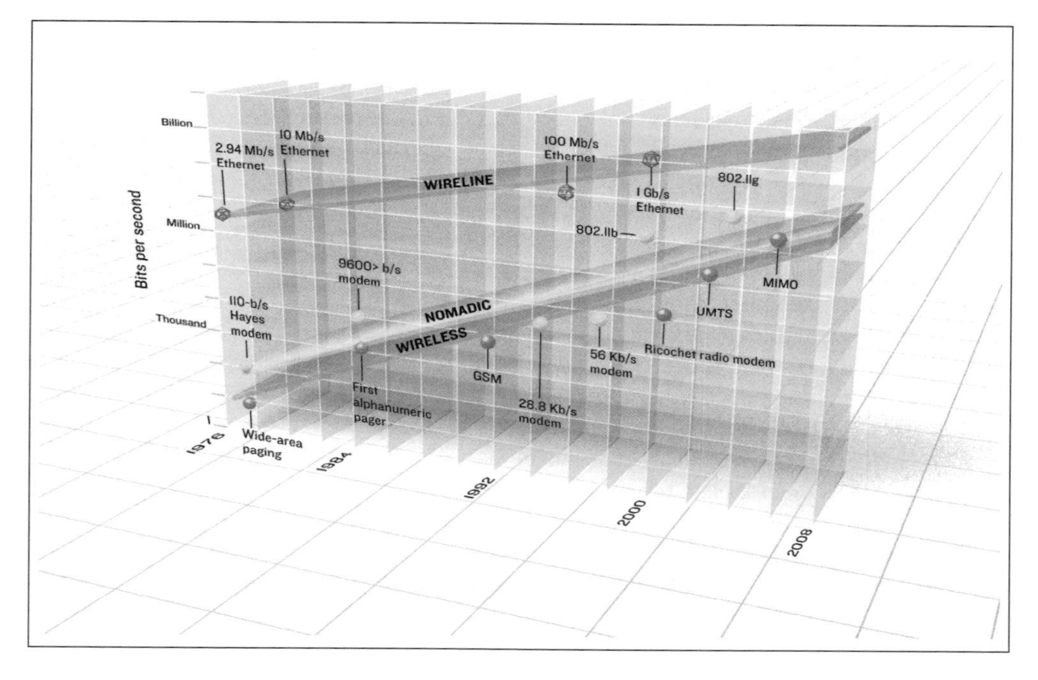

Figure 1.6 Edholm's Law: fixed and wireless transmission rate trends. Reproduced by permission of IEEE Spectrum Magazine, July 2004

Inc. chief technology officer, this figure shows that wireline (fixed), Nomadic (wireless without mobility), and wireless (mobile) communications have been growing 'almost in lock step'. The rate of growth for wireless (mobile and nomadic) communications is faster than that of wireline communications. Extrapolating forward, it appears that we may someday see a convergence between wireline and wireless rates. If the cost of wireline infrastructure remains higher than that of wireless, as it is today, then the end of wireline communications may very well arrive.

There are two desirable aspects to internet connection: fast transmission rates and seamless, constant connectivity. Fast transmission rates require wide frequency bandwidths, as well as high bandwidth utilisation efficiency. Wider bandwidths must be decided by governments and agreed upon internationally, a subject we will discuss more in the next chapter. Discussions on possible bands for broadband wireless systems are already under way in many parts of the world, as listed in Figure 1.7. As increasingly wider bandwidth is required for future systems, it is inevitable that higher frequency bands should be used. There are however, advantages as well as disadvantages to operating in higher bands. We will discuss these in the next chapter. While in fixed-line communications between two users the whole medium is dedicated for their connectivity, the available bandwidth for wireless communications is limited and must be shared among many users. The challenge of all wireless technologies is how to use this common resource efficiently. Moreover, the resource must be allocated in such a way that each user *experiences* uninterrupted connectivity.

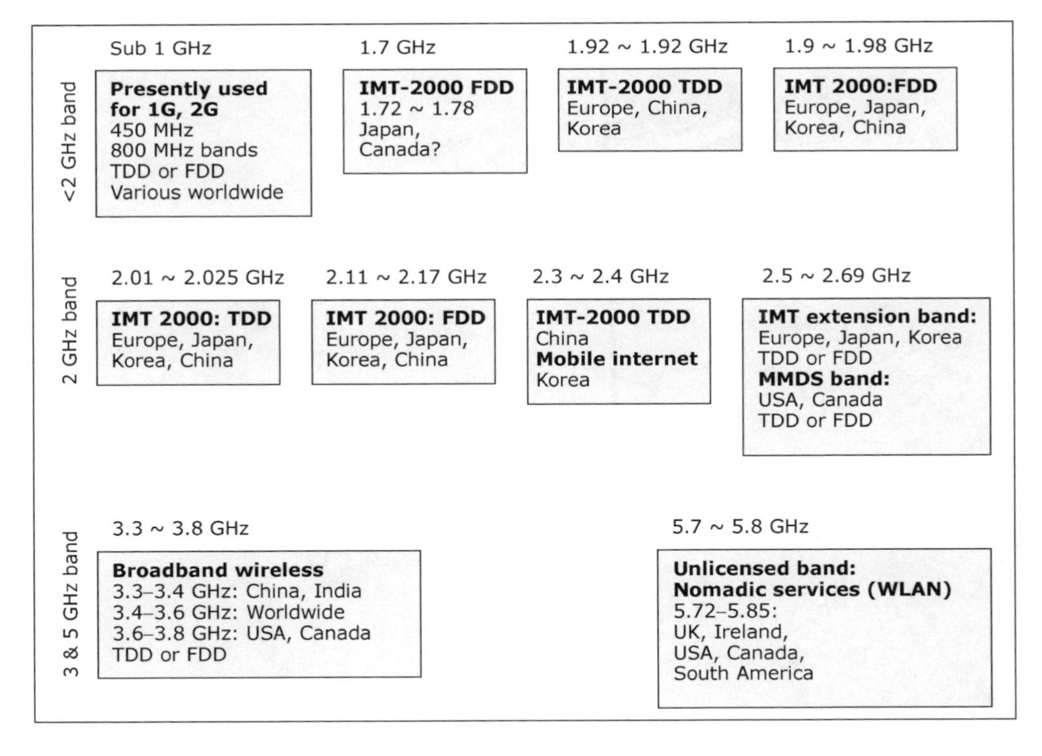

Figure 1.7 Possible spectrum for broadband wireless

Research and development for the next-generation systems, sometimes referred to as 4G, are following two different paths. One is an evolutionary track from the present 3G systems, sometimes referred to as super 3G. The research is focused on adapting the present standards to full data communications, and includes technologies to increase transmission rates. The second is a revolutionary track, proposing new radio access technologies designed specifically for broadband wireless communications.

1.5 Duplex Modes

Duplex communications between a central BS and end-user equipment can be carried in two modes. In the first mode, two separate frequency bands are used for downlink (DL), or BS to end-user, and uplink (UL), or end-user to BS transmissions. This is called *frequency division duplexing* (*FDD*). FDD mode is used in most 1G and 2G systems as it is well suited for voice communications. In the other duplex mode, UL and DL transmissions are carried out in the same frequency band. The band, however, is alternately switched for BS and end-user transmissions. This is called *time division duplexing* (*TDD*) and has been used in some short-range 2G systems such as Personal Handyphone Systems (PHS) and Digital Enhanced Cordless Technologies (DECT). For 3G systems, two FDD standards Wideband CDMA (WCDMA and CDMA2000) and two TDD standards (Time Division CDMA – TD-CDMA and Time Division – Synchronous CDMA – TD-SCDMA) have been defined. At this point in time, it is unclear whether 4G systems will be based on TDD or FDD. These two duplexing systems, FDD and TDD, are illustrated in Figure 1.8.

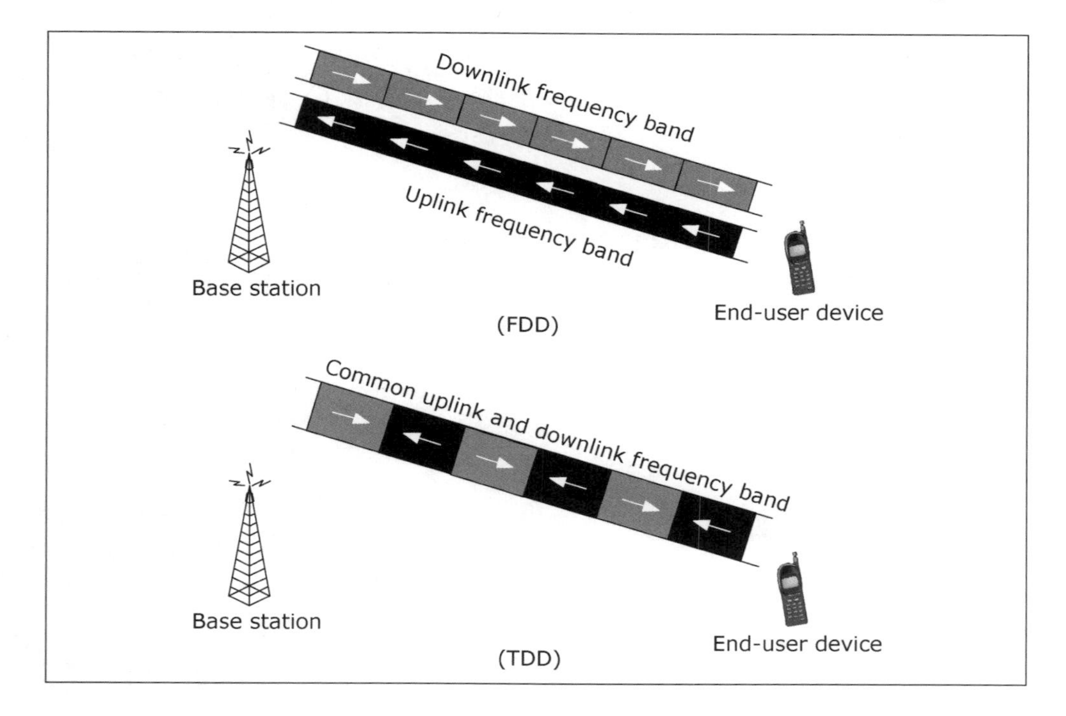

Figure 1.8 FDD and TDD modes

Reciprocity

A major characteristic of TDD systems is channel reciprocity. The performance of radio communications systems is highly dependent on transmission channel characteristics. Because the channel is highly time variant, equalising for channel variations is very important. Since channel variations depend on the frequency band, channel variation characteristics are uncorrelated for the UL and DL of FDD systems, which use two separate bands for DL and UL. For this reason, feedback processes are required for channel equalisation in FDD systems. However, for TDD systems, since the same frequency is used for both DL/UL transmissions, channel variations are highly correlated. This is known as *TDD channel reciprocity* because variations between UL and DL are reciprocal. This means TDD systems do not require feedback for most of their channel equalisation purposes.

Resource allocation

DL and UL resource allocation is an issue of great importance to broadband wireless systems. Voice communications require equivalent resources for UL and DL. For data communications, however, traffic amounts are not necessarily equivalent. That the traffic volumes for DL and UL vary in an unexpected fashion is a phenomenon we will discuss below. What resources ratio will be required for each link in the future broadband wireless systems remains uncertain.

A major advantage of TDD over FDD is its flexibility of resource allocation. As capacity allocation can be carried by allocating a portion of time to each link, reallocation can simply be done by moving the time switch, or changing slot allocation ratios as illustrated in Figure 1.9. In fact, capacity allocation can be varied as often as desired, and even independently at the cell level. In contrast, it is nearly impossible to reallocate an FDD spectrum once it has been decided. This is one reason TDD systems are considered to be a stronger candidate for broadband wireless systems of the future. We will illustrate this effect using an example, in Chapter 9.

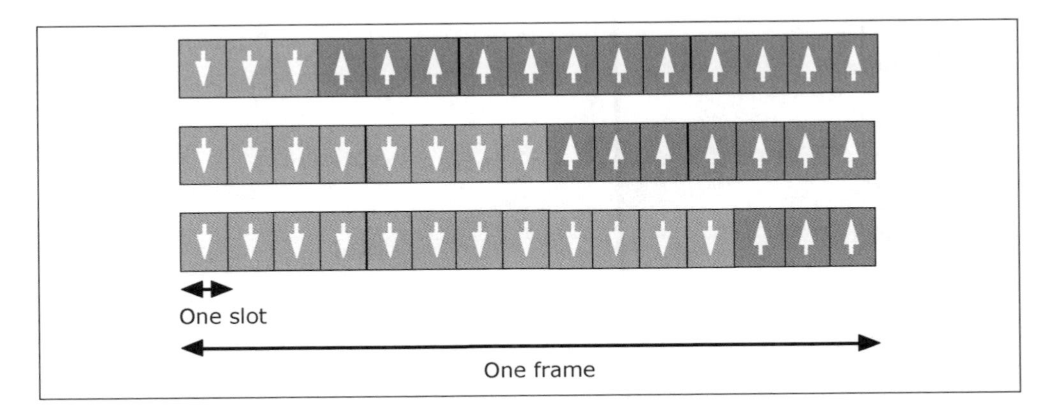

Figure 1.9 Downlink/Uplink slot allocation in TD-CDMA standard: one frame is 10 ms and is divided into 15 slots that can be allocated to either downlink or uplink

1.6 Voice to Data

For almost the entire first century of telecommunications history, voice had been by far the primary service provided over fixed links. In contract, over the past two decades, it is fixed-link data communications traffic volumes that have been growing. As shown in Figure 1.10, they have now surpassed the voice traffic volume.

The same is expected to happen with wireless data and voice traffic. The transition from a voice-centric service, as 1G and 2G systems have been, to what will be data-centric services is one of the most significant aspects of system design and operation for 3G and post-3G systems. Voice communications generally require connectivity for the duration of a call, and resources are allocated on a continuous basis. This is known as *circuit switching*. The technology dates back to telephony days when circuits were switched, manually, for each voice call. Data transmissions, however, do not require continuous allocation: resources can be allocated on per-packet basis. This is known as *packet switching*.

1.6.1 Voice-over internet protocol

With the emergence of full packet-switched networks, voice communications, both fixed and wireless are also going through a revolution. In fixed-line communications, Voice-over IP (VoIP) is fast replacing old circuit-switched voice networks. It is expected that by 2008–2009 the bulk of voice traffic will be carried using VoIP technology.

Wireless, mobile voice communications of 1G, 2G, and 3G still use circuit switching. The more recent 3G systems, however, are standardised to use an all-IP backbone and packet

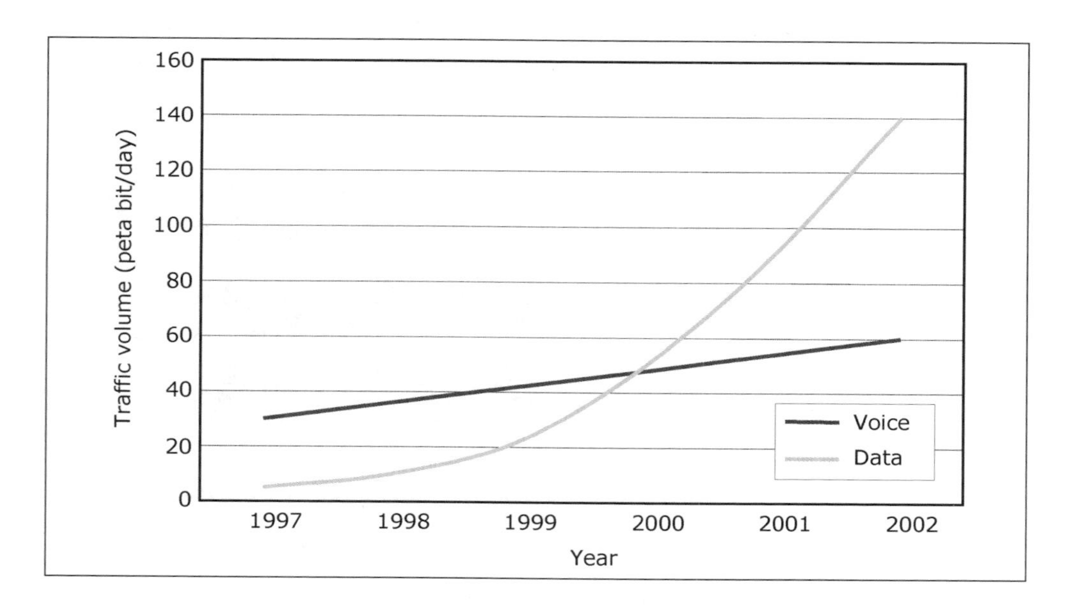

Figure 1.10 Voice and data traffic. Reproduced by permission of IEEE

Table 1.5 Packet switching in 3G

Service	1G and 2G	3GPP Release 99	3GPP Release 5 and beyond
Voice	Circuit switched	Circuit switched	Packet switched
Video	—	Circuit switched	Packet switched
SMS	Circuit switched	Packet switched	Packet switched
WAP	Circuit switched	Packet switched	Packet switched
E-mail	Circuit switched	Packet switched	Packet switched
Web	—	Packet switched	Packet switched
MMS	—	Packet switched	Packet switched
Streaming	—	Packet switched	Packet switched

switching as shown in Table 1.5. However, transmission of VoIP packets over the air is not efficient as packet overheads are mostly unnecessary. Quality of service requirements need to be considered in air-interface design for VoIP services. We will discuss this issue further in the next chapter.

1.7 Traffic Profiles

Another issue affecting business decisions in the development of the next-generation wireless communications system is the user's service requirements. Both 1G and 2G systems were designed for voice communications. Voice traffic characteristics had been studied for over a century and were well known. It was easy to calculate the required number of voice channels based on population density and market expectations. Moreover, the DL and UL traffic characteristics were highly similar. FDD systems are well suited to this kind of traffic (symmetric up and DL capacities), with the precedent that the public, fixed-line telephone systems also used an FDD mode of operation.

In contrast, the next-generation broadband wireless systems are mostly for data transmission applications such as sending and receiving files and browsing the internet. Models for this kind of traffic have been developed only recently. In particular, web browsing had been considered the major traffic activity on the internet, where a user accesses a web site using a short data burst, downloads a rather large-sized web page, and spends a certain time reading the downloaded information.[1] It was therefore believed that the DL traffic is comparatively larger than the UL, implying that spectrum allocation should be larger for the DL.

However, a recent boom in peer-to-peer (P2P) file exchange, particularly audio and video files, is expected to balance the UL and DL traffic volumes for personal communications. Napster, Kazaa, and other file sharing P2P programs (see Box: File sharing programs) well represent this trend. In fact, Kazaa is currently the most downloaded software. The following observations on P2P traffic in 2004 in Europe signify this trend:

- P2P file sharing accounts for over 70% of the traffic on ISP networks.
- P2P traffic volume accounts for 95% of uplink traffic over networks' last mile.

[1]UMTS document TR 101 112 defines a web-browsing model for a WWW browsing session.

File-sharing software

Sharing files over the internet has become easily possible using any one of a number of freely available file-sharing programs, such as Napster, KaZaA, eDonkey, iMesh, Lime Wire, and so on. Among these, KaZaA has the reputation for being the most downloaded software program (389 million downloads and counting).

- An average of five million people were connected to a P2P network.
- Thirty-five million people in Europe had downloaded music from a P2P network.

How long will this trend last? And what does this fixed-link communication trend mean to wireless communications? Does this mean that the same amount of spectrum must be allocated to UL and DL? There are no models yet defined for this kind of traffic for wireless, and it is still too early to accurately indicate how DL and UL traffic patterns will develop, even in general, for the next generation of broadband wireless systems. A flexible design that allows for dynamic reallocation of capacity to UL and DL will prove valuable.

1.8 Access Technologies

Several access technologies have been proposed for wireless broadband. Some of these technologies are presently being used in 2G and 3G systems. A summary of the present technologies and their applications for broadband wireless is presented in the following sections. Table 1.6 shows the duplex mode and carrier bandwidth specifications for several standards.

1.8.1 Frequency division multiple access

Frequency division multiple access (FDMA) is the oldest of the multiple access technologies, and was primarily used in the analogue 1G systems. In FDMA, the total available bandwidth is divided into a large number of small carriers, each of which is used to

Table 1.6 Carrier bandwidths of several standards

	AMPS	GSM	WCDMA	TD-CDMA	WiMAX
Duplex mode	FDD	FDD	FDD	TDD	FDD & TDD
single channel bandwidth	30 kHz	200 kHz	5 MHz	5, 10 MHz	1.25 $\sim$ 20 MHz

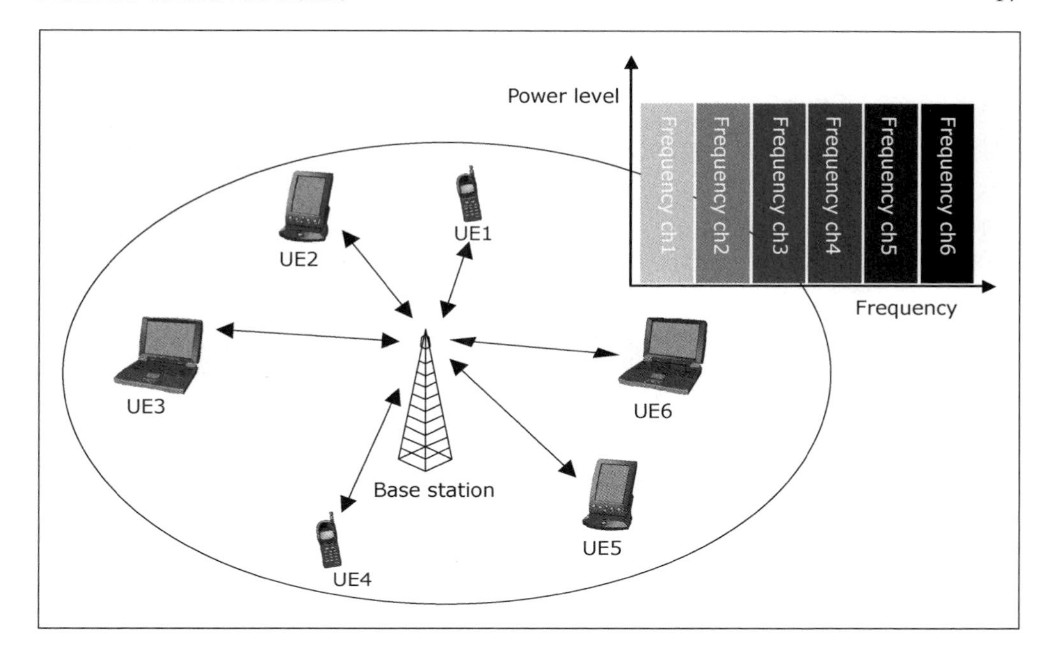

Figure 1.11 Frequency division multiple access

carry voice traffic from a base station to a mobile device (and vice versa), as illustrated in Figure 1.11. FDMA technology, in its initial form, was abandoned in favour of more advanced, more efficient digital technologies in 2G systems. Frequency division multiplexing has appeared again in a more advanced format, as will be discussed in Chapter 2.

1.8.2 Time division multiple access

In Time Division Multiple Access (TDMA) technologies, several users access the central base station using the same frequency band, but not all at the same time. Each of them takes turns to connect to the base station one at a time as shown in Figure 1.12. This technology is used in most 2G systems, such as Digital AMPS (D-AMPS), PDC, and GSM.

1.8.3 Code division multiple access

The CDMA technology (Figure 1.13) became popular thanks mainly to the pioneering work of Qualcomm Inc. The technology is based on the spread-spectrum technique, known and well practised in military communications for over 50 years. In one form of spread spectrum, the signal is spread over a wide bandwidth through multiplication with a pseudo-random code. The signal can then be detected by similar multiplication by exactly the same code. CDMA systems work on the principle that two or more users' signals may be transmitted, and distinctly received, in the same band as long as they use distinct spreading codes for their signals. After spreading, all signals occupy the same bandwidth and appear

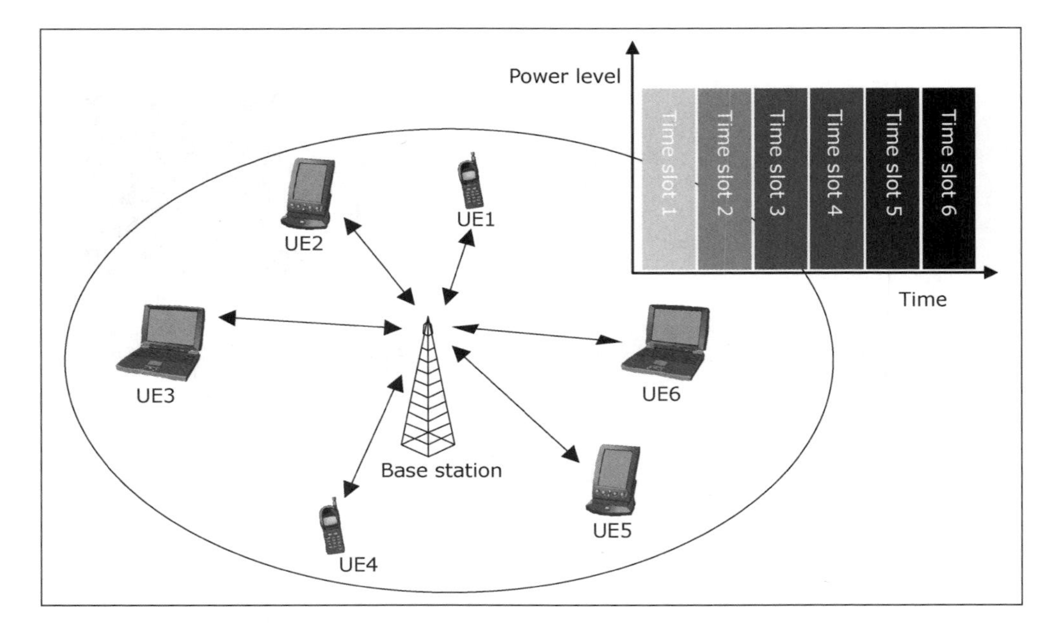

Figure 1.12 Time division multiple access

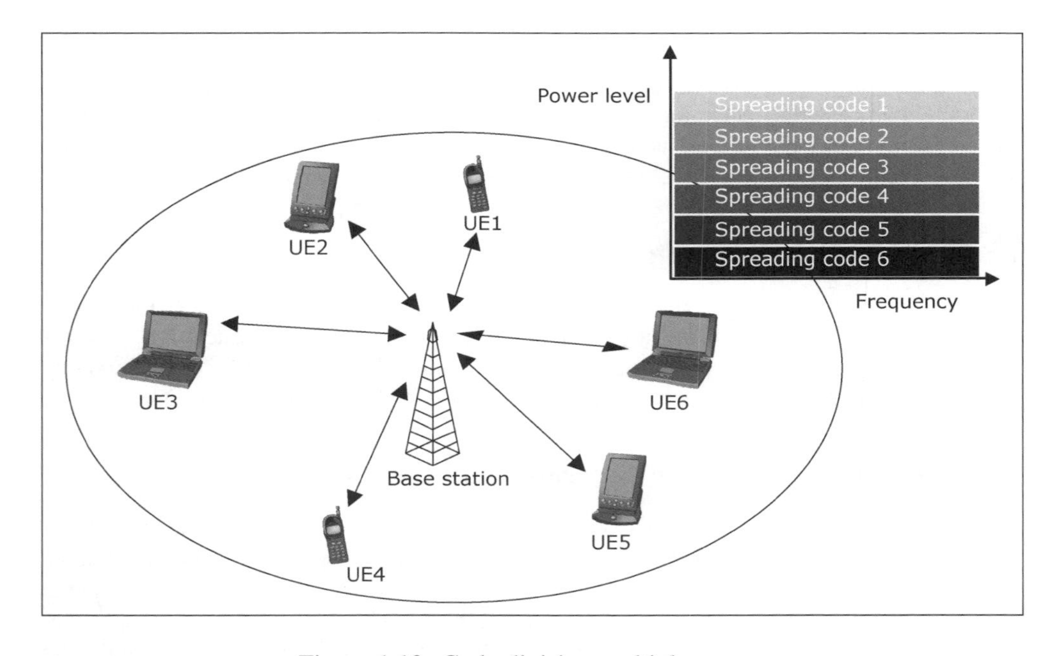

Figure 1.13 Code division multiple access

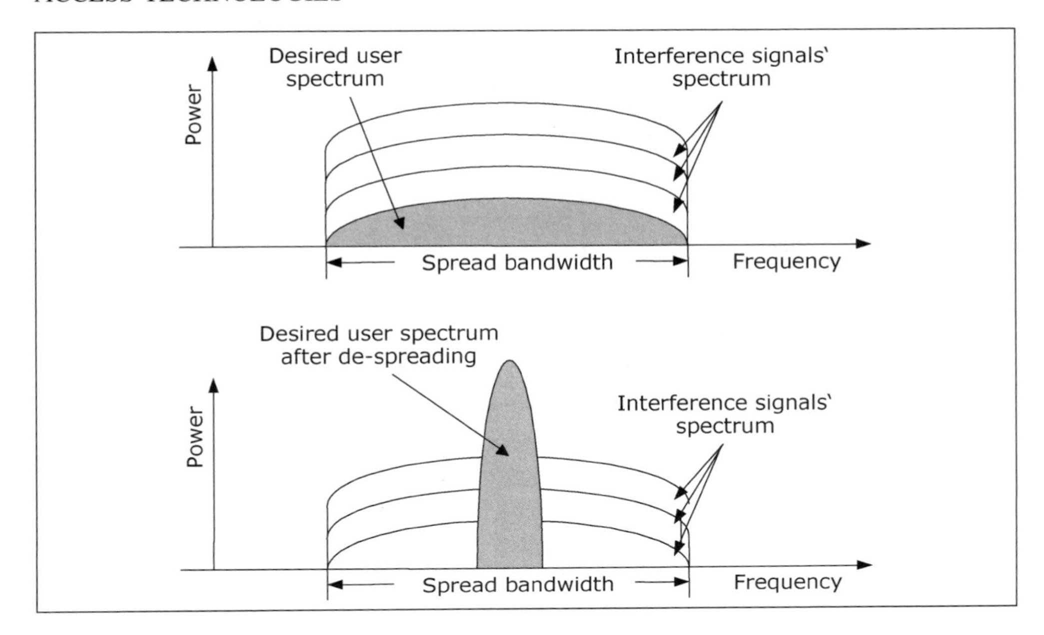

Figure 1.14 Wide-band spread signals and narrow-band desired signal after de-spreading

to each other as noise. At the receiver side, a de-spreading function is carried out, which restores only the desired signal to the original narrow band, as shown in Figure 1.14.

CDMA was used in the IS-95 standard, and competed directly with TDMA systems such as D-AMPS and GSM for the 2G market share. CDMA systems can be said to form the basis for all 3G systems. Although some evolutions of 2G TDMA systems have been considered for 3G, the mainstream systems are based on CDMA. There are two leading sets of 3G standards: one comes from the third-generation partnership project (3GPP) and presently has three modes: WCDMA, TD-CDMA and TD-SCDMA. The other 3G standard has been developed by 3GPP-2, and is an evolution of the 2G IS-95 standard.

1.8.4 Channel sense multiple access

Wireless LAN (WLAN) network operations are based on a self-organising decentralised transmission control topology. In WLANs, each user monitors the carrier transmission channel to detect whether any other user is transmitting. The process is to carry request-to-send (RTS) and clear-to-send (CTS) signalling between two users. This is known as *channel sensing*, and this mode of channel access is known as *Channel Sense Multiple Access* (*CSMA*). If no other device is using the channel, then the user equipment proceeds to transmit its message.

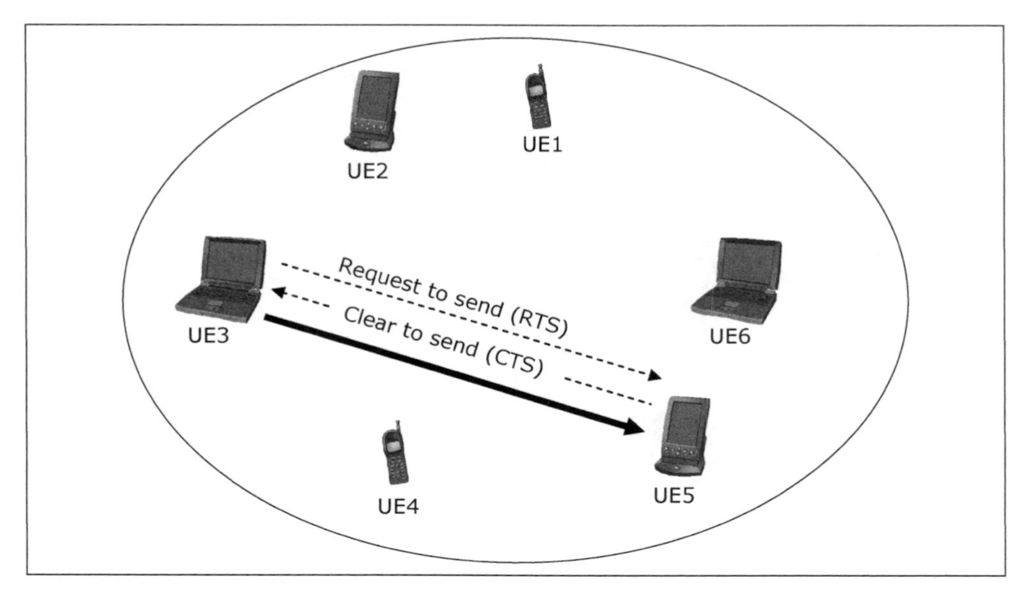

Figure 1.15 Channel sense multiple access

Table 1.7 Communications technologies and their multiple access domains

Communication technologies	Multiple access domain
AMPS	Frequency
GSM	Time and frequency
WCDMA	Code
TD-CDMA	Code and time
WLAN	CSMA + frequency/code and time
WiMAX	Frequency and time

Collisions may occur if two users decide to monitor the channel at the same time and, upon detecting that it is free, proceed to transmit. Collision avoidance algorithms are utilised to reduce the probability of occurrence of such an event. Figure 1.15 illustrates the operation of CSMA systems.

Wireless LAN system have been standardised through organisations such as IEEE and ETSI (see Glossary). Several WLAN standards, and versions, exist, where each uses a specific combination of CSMA and modulation, coding, and antenna technologies. In general, present mobile and nomadic technologies use one or a combination of the above access techniques. Table 1.7 summarises examples of present communication technologies and the multiple access methods they use.

1.9 Telecommunications Operator Business

The telecommunications operator business can generally be described as a fixed-cost business. The initial costs for designing and construction of a network are quite high. In comparison, operating costs per single subscriber are small. Similar to other fixed-cost businesses, airlines, for example, an operator needs to attract as many customers as possible to the service in order to recover the initial costs, and to operate profitably.

With (virtually) unlimited capacity, and in the presence of competition, costs to end users will shrink to leave the operator with only marginal profits. This is what has happened to long-distance telephone businesses. Many such carriers have faced financial troubles, and have been trying to find new revenue-generating activities. Some have filed for bankruptcy protection, and some have even ceased to exist as independent companies. A few prominent examples are shown in the box 'Long-distance operators in trouble'. However, when capacity is limited, the operators can maintain a healthy profit margin. This is because competitors cannot add more subscribers when their networks reach full capacity, and therefore the competition for attracting new subscribers is not very fierce. A look at the ARPU figures for different countries provides a good indication of the demand and supply of mobile telephone capacity. Table 1.8 shows the ARPU figures in 2004 for several countries for subscriber contracts (excluding prepaid ARPU). Of course, other factors such as a lack of (real) competition as well as government regulation also determine the end-user costs.

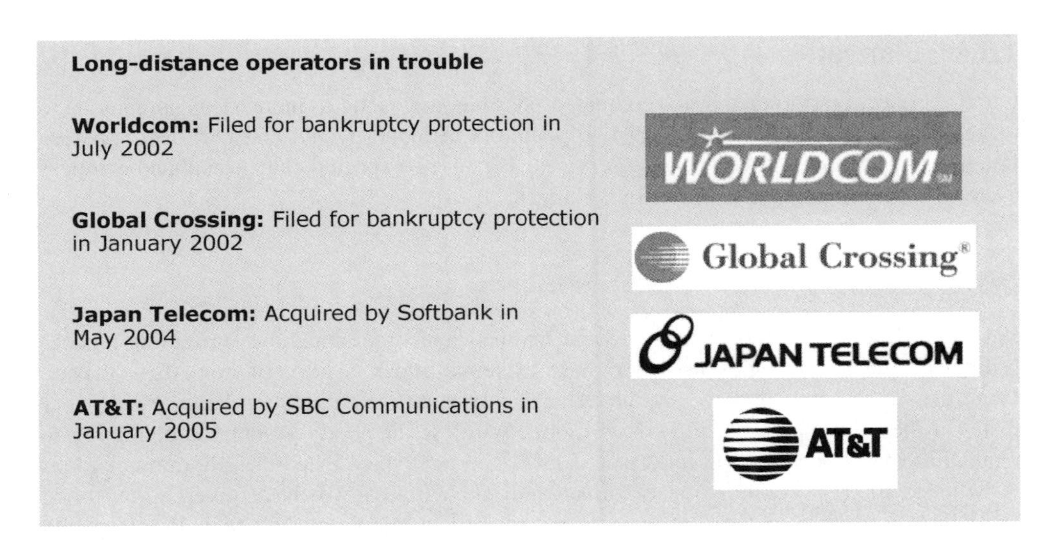

As governments release more spectrum for wireless communications, the danger of ARPU meltdown to existing operators becomes more and more real. Will wireless communications go the way of long-distance telephones? Will these operators also become the supplier of a transmission pipe and no more?

Table 1.8 ARPU figures for operators in several countries

Country	Operator	ARPU (US$)
Australia	Vodafone	55.47
Singapore	StarHub	42.36
Japan	DoCoMo	73.16
UK	T-Mobile	69.51
South Africa	Vodacom	83.93
USA	Cingular	50.32
Argentina	Nextel	40.00

1.9.1 From pipe to content provider

Two strategies are being followed by wireless operators. One is to increase the traffic per subscriber, and thereby keep traffic demand equal to possible supply. The increase in voice traffic responds to lower costs, but is limited by size of population. The other is to rise in the value chain, and engage in providing content to the end-user. The operator could become an outright content provider or at least a stakeholder in providing content. An example is the i-mode model of Japan, as shown in Box 'Japanese i-mode service' on page 8.

1.9.2 Flat rate

Once operators start to charge for content, a possible new tactic to increase usage may be to offer flat-rate subscriptions. Already two operators in Japan, KDDI and DoCoMo, provide flat-rate subscriptions for data-only services. It can be expected that broadband wireless systems will be operated with flat-rate charges.

Summary

We have made a review of broadband and wireless telecommunications, providing a background of the technologies used for these systems, and a history of how they arrived. From a market point of view, the broadband data communications field is growing very fast. On the other hand, the growth of mobile/wireless telephony systems is beginning to plateau as market penetration rates near 100%. The next stage in wireless communications is believed to be a combination of broadband and wireless. We have given a review of technologies used in successive generations of mobile telephony and their user capacity characteristics.

In relation to broadband wireless technology, we have discussed the characteristics of the traffic likely to be carried over these networks. Furthermore, we have briefly discussed how a broadband wireless operator may choose to do business. In the following chapters, we will go into further detail on the topics discussed here.

Further Reading

- On ISDN:
  - "ISDN Basic user-network interface – Layer 1 specification" International Telecommunication Union-Telecommunications Standards Section (ITU-T) I.430.
- On DSL:
  - Technical and marketing reports from DSL Forum: `http://www.dslforum.org/`, 2005
- Statistics on mobile market development in Japan:
  - Yearly white paper by the ministry of home affairs and communication: `http://www.johotsusintokei.soumu.go.jp/whitepaper/eng/WP2005/2005-index.html`
- On i-mode:
  - Natsuno, T., *The I-Mode Wireless Ecosystem*, John Wiley & Sons, 2003a.
  - Natsuno, T., *I-Mode Strategy*, Halsted Pr, 2003b.
- On WAP:
  - WAP forum web site: `http://www.wapforum.org/`, 2005
- 3G standards:
  - 3GPP specifications: `http://www.3gpp.org/`, 2005
  - 3GPP2 specifications: `http://www.3gpp2.org/`, 2005
  - Holma, H., Toskala, A. (Editors),*WCDMA for UMTS: Radio Access for Third Generation Mobile Communications, 3rd Edition*, John Wiley & Sons, 2004.
- On IEEE Standards:
  - IEEE 802.11 family, (WLAN Standards): `http://grouper.ieee.org/groups/802/11/index.html`, 2005
  - IEEE 802.16 family, (includes WiMAX standard): `http://grouper.ieee.org/groups/802/16/index.html`, 2005
- On Wireless technologies:
  - Pahlavan, K., Levesque, A. H., *Wireless Information Networks*, John Wiley & Sons, 1995.
  - Rappaport, T. S., *Wireless Communications: Principles and Practice, 2nd Edition)*, Prentice Hall, 2001.

2

Wireless Communications

This chapter gives a brief technical background of the issues involved in communicating over the frequencies of interest to broadband wireless. First, we discuss the characteristics of signal fading in the radio channel and how access technologies deal with this fading. An important aspect of the access technologies mentioned in the last chapter is how effectively they operate under, and compensate for, fading conditions. Next, we discuss the issue of equalisation with regard to co-channel, multi-user interference. The difference between these technologies is almost philosophical, that is, how to work in a difficult communications channel. This philosophical choice does lead to differences in performance, operational simplicity, and ultimately, cost. We then describe diversity communication, which is another method to combat channel fading. With this, we will compare the performance of various access technologies in terms of the transmission rates they support.

In this chapter, we also describe the shift from voice to data communications, how the new systems operate, and what this means for access technologies. We then describe the quality of service (QoS) standards in relation to services that may be provided over broadband wireless access systems. This chapter is intended to provide answers to 'why' the wireless telecommunications technologies have evolved to their present state.

A simple block diagram of a wireless communications system is shown in Figure 2.1. The information data bits are encoded, modulated, and then transmitted over the air. Before and after transmission, equalisation processes are carried out to ensure that a certain degree of transmission quality is maintained. Demodulation followed by decoding processes yield replicas of the transmitted data bits. We will discuss about these blocks in more detail in the following text.

2.1 Signal Fading

First, let us discuss the characteristics of a wireless transmission channel. The transmission and access technologies mentioned in the last chapter have been designed to operate in the constantly varying channel conditions associated with 2∼5 GHz frequency band mobile communications systems. Generally, in fixed-line channels, such as those associated with ISDN or DSL technologies, the received power level is long-term time invariant, that

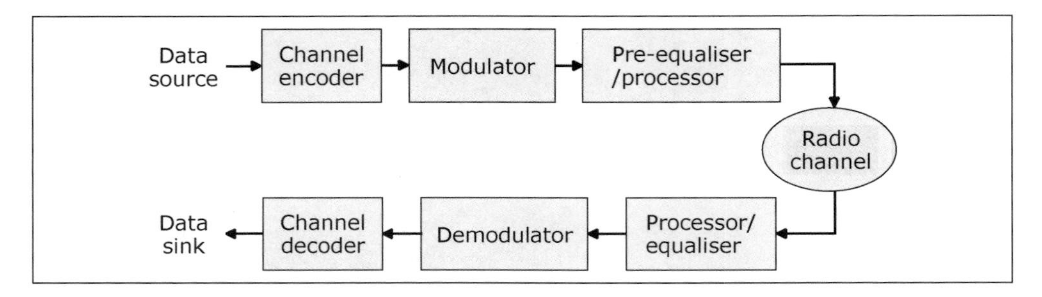

Figure 2.1 A wireless communications system

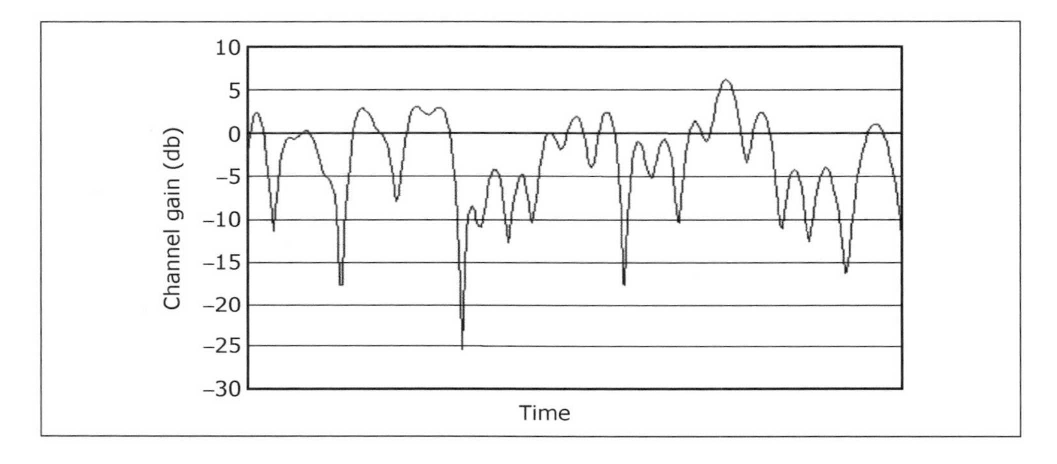

Figure 2.2 A flat fading pattern

is, broadly constant. Signal degradation in such fixed-line channels varies mostly from thermal noise and cross talk. As individual insulated copper-wire pairs from subscribers lay alongside each other, signals at high operating frequencies cross from one wire to the next, causing interference. To these is added frequency-specific losses associated with copper-wire transmission lines. Even so, the received power level is mostly a function of the length of the link, which is the distance between the home and the exchange. In contrast, in addition to loss due to distance, mobile communication channels are also time variant, that is, the received power level changes rapidly over time. The changes in power level can be quite large and of several degrees of magnitude. This phenomenon is referred to as *signal fading*, and is illustrated in Figure 2.2. Signal fading occurs in both the time and frequency domains. Also, path loss, as a measure of signal attenuation versus distance, is a function of operating frequency and it increases as higher and higher bands are used for communication.

2.1.1 Why fading?

Why do received signals fade? Typically, radio communication channels are characterised by multi-path signal arrival. The received signal is composed of many reflections of the

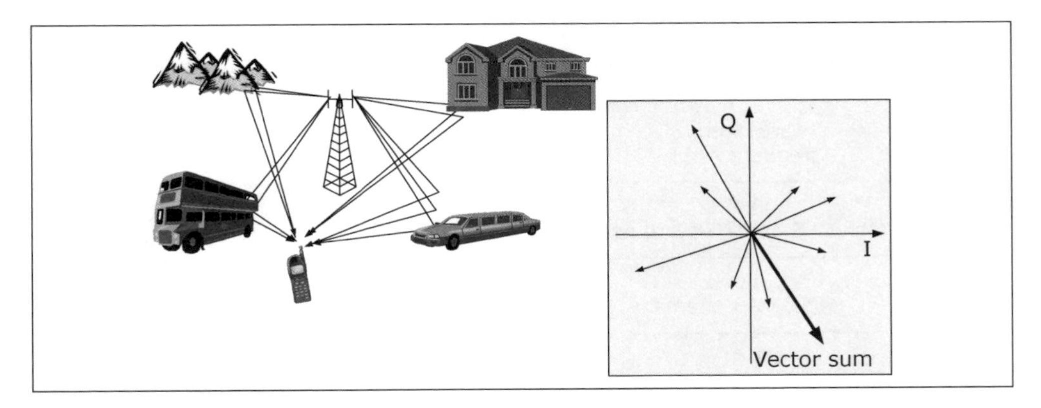

Figure 2.3 Multi-path signal arrivals and their vector summation

transmitted signal – reflections from natural and man-made objects such as mountains, buildings, cars, and the ground, and so on (as illustrated in Figure 2.3). As all of the reflected signals travel through different paths, they have different propagation delay times, amplitudes, and phases. At each instant and at each physical point, these signals may combine constructively or destructively. This phenomenon is illustrated by the vector representation of signals arriving at a receiver antenna in Figure 2.3. Each ray of a received signal is a vector with a certain phase and amplitude. If received signals are drawn for an area, the sum of the vectors will appear as peaks and troughs – peaks representing the places where multi-path signals combine constructively and troughs indicating where they combine destructively.

An end-user receiver moving in this environment experiences periods of good and bad reception, with the peak-to-trough ratio at times exceeding 50 dB. The variation in reception level is known as the *fading rate* or the *Doppler rate*, f_d. Its value depends on the frequency of operation (carrier frequency f_c) and the velocity (speed V) of the receiver's movement with respect to the transmitter. A rule of thumb for calculating the Doppler rate is as follows:

$$f_d \approx f_c \times V$$

where f_d is expressed in hertz (Hz), f_c is expressed in gigahertz (GHz), and V is expressed in km/h. Sample Doppler rates for several broadband wireless access systems have been calculated and are shown in Box 'Examples of Doppler rates'.

A typical fading pattern is shown in Figure 2.2. The physical meaning of the Doppler rate is the number of times the signal level crosses a certain threshold per second. This fading pattern is typical of narrow-band systems, which have signal bandwidths in the order of a few hundreds of kHz. This kind of fading is generally referred to as *flat fading*. To illustrate the effect of operating in such a channel, we show the bit error rate (BER) performance (see Box: Bit error rate) for two channels in Figure 2.4. One is the above-described flat-fading channel and the other is a constant channel with no fading. We can observe that the average received (or corresponding transmission) signal power needs to be more by 18 dB (63 times) in a fading channel as compared with a constant channel to yield a similar performance quality of BER equal to 10^{-3}.

Examples of Doppler rates

Operating frequency (f_c)	Travelling velocity (V)	Doppler rate (f_d)
800 MHz (Most 1G and 2G systems)	100 km/h	80 Hz
2.0 GHz (Most 3G systems)	100 km/h	200 Hz
5.8 GHz (Some beyond-3G systems)	100 km/h	580 Hz

Bit error rate

Bit error rate (BER) and its related packet error rate (PER) are common communications quality metrics, and are defined as the ratio of the number of erroneously received bits (packets) and the total transmitted bits (packets). As received signal quality (with respect to noise) improves, the number of errors decrease. It is therefore appropriate that BER/PER are expressed with respect to signal-to-noise ratio (SNR) or a similar parameter. Generally, the larger the SNR, the lower the BER/PER.

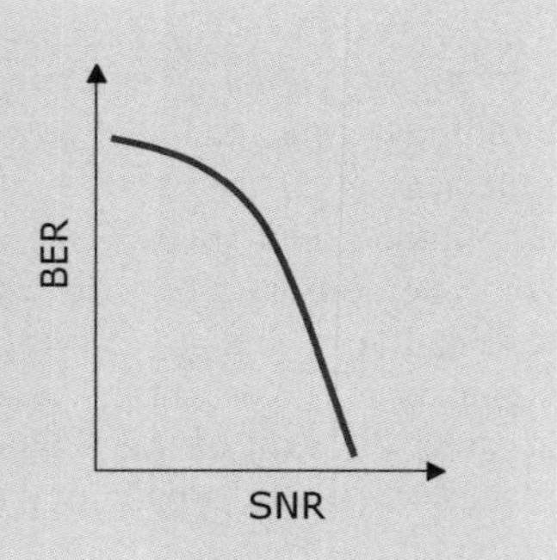

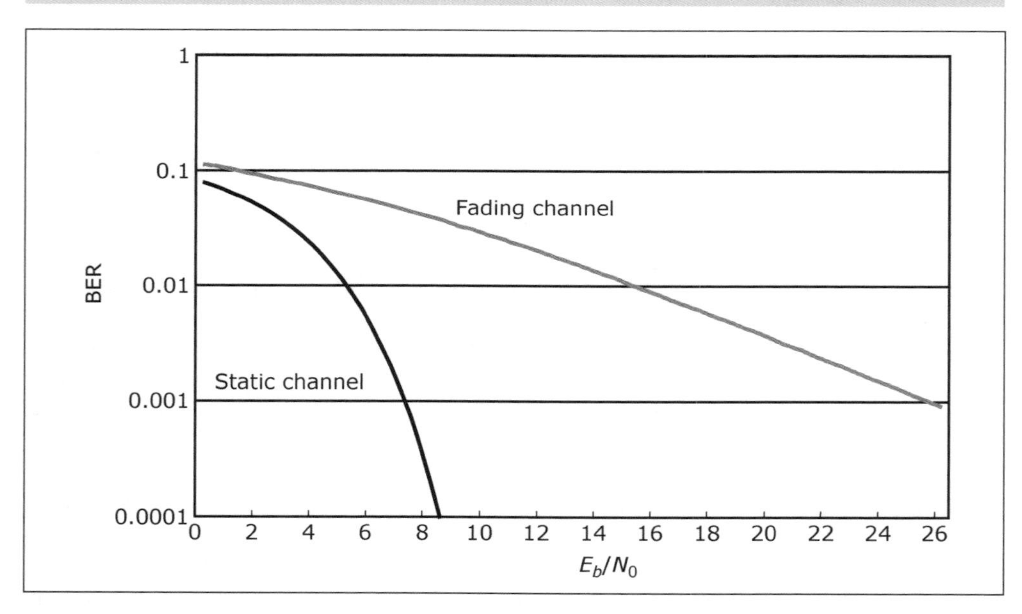

Figure 2.4 Bit error rates (BER) in flat-fading and constant channels

Flat fading, as described above, is representative of narrow-band systems. Examples of these are 1G and 2G systems, with the exception of CDMA-based IS-95. In systems with a wider frequency band, typically more than 1 MHz of bandwidth, fading behaviour is no longer flat. In these systems, changes in signal level in the frequency domain differ for different parts of the signal spectrum. This phenomenon is known as *frequency-selective fading*, where some components of the frequency fade while other parts remain unfaded.

Frequency-selective fading occurs owing to large propagation delay differences between arriving multi-paths. A measure of this delay spread is the difference between the arrival of the first and last significant multi-path signal. This is known as *maximum delay spread*. Here, a significant multi-path signal is one that is larger than a certain threshold compared with the total received signal power.

A measure of frequency-selective fading is channel coherence bandwidth. Coherence bandwidth is defined as the maximum frequency bandwidth over which the fading changes are correlated for all frequencies of the band. This is inversely proportional to the channel's maximum delay spread, that is, the more the channel's delay spread, the less the coherence bandwidth and therefore the more the frequency-selective fading. The behaviour of a frequency-selective fading channel in both time and frequency domains is shown in Figure 2.5.

An example of delay spread is shown in Figure 2.6. Note that since delay spread is the result of the communication channel's physical features, frequency-selective fading is independent of the communications technology. It is merely a result of radio transmission in a physical environment. A transmission channel in a hilly terrain with many buildings may appear to have highly frequency-selective fading because of large delay spreads, whereas a channel in a plain with no buildings would have a very low delay spread and flat fading.

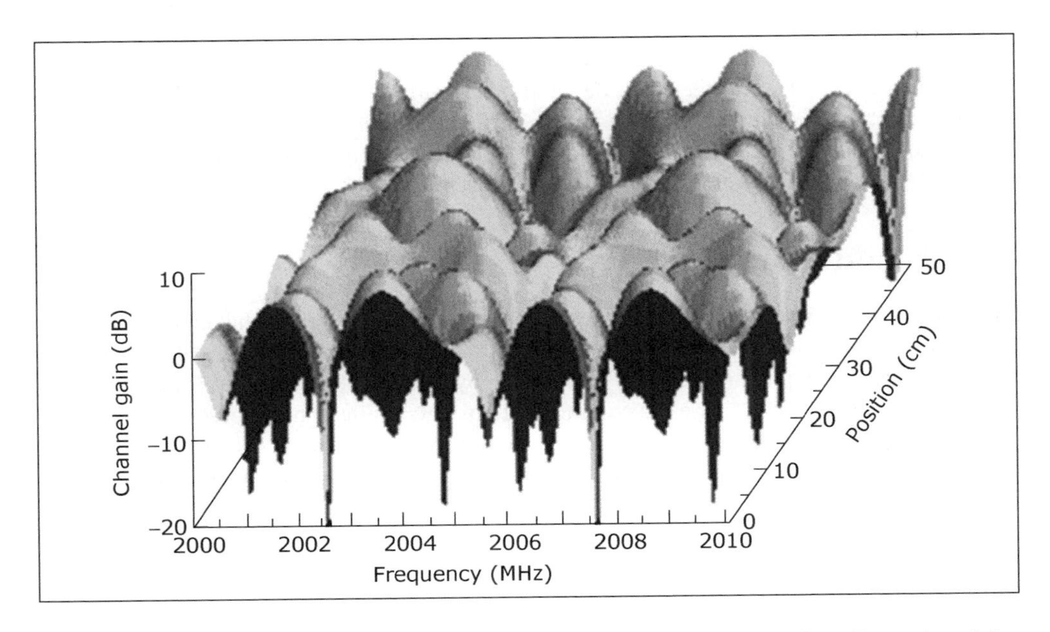

Figure 2.5 Frequency-selective fading in time and frequency domains. Reproduced by permission of Professor Fumiyuki Adachi

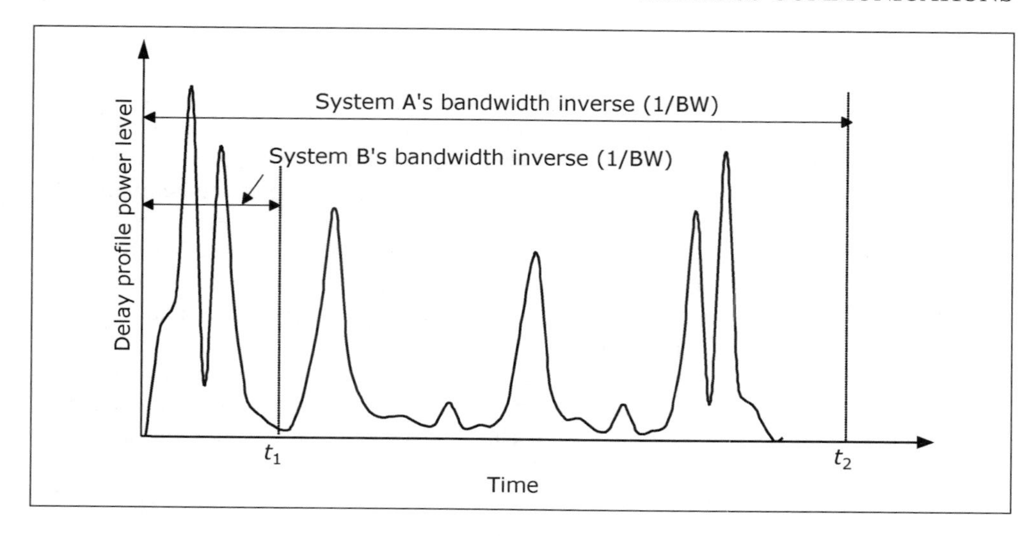

Figure 2.6 Delay spread

Wireless systems experience flat fading or frequency-selective fading depending on their transmission bandwidth. If the bandwidth is larger than the coherence bandwidth, fading is frequency selective, and if smaller it is flat. In Figure 2.6, two systems have been illustrated. In system A, channel fading is flat, whereas system B exhibits frequency-selective fading behaviour. For a typical urban maximum delay spread of a few microseconds, the coherence bandwidth is in the order of 500 kHz. For such a channel, all 2G systems, excluding IS-95, experience flat fading. In contrast, all 3G and 4G broadband wireless systems require high transmission rate, and therefore demand bandwidths of several MHz. Therefore, in a majority of channel environments, the fading is frequency selective. Some of these systems employ multi-carrier transmission techniques, which ensure that fading on each sub-carrier becomes flat. These technologies will be discussed in Chapter 3.

2.2 Modulation

Modulation is the process of converting a series of data bits to signals that can be transmitted over a communications medium. As transmitted signals constitute an electromagnetic waveform, signal modulation can be carried out by altering the magnitude, frequency, or phase of this waveform. In wireless communications, a combination of phase and amplitude modulation is considered most efficient in terms of system capacity. At the receiver side, a coherent (i.e. phase locked) detection of the received signal is performed to reconstitute the transmitted data.

2.2.1 Signal constellation

Two common modulation schemes used in wireless communications are multi-phase shift keying (m-PSK) and multi-quadrature amplitude modulation (m-QAM). For example, a 2-PSK, or (bi-phase) BPSK system maps the incoming bit stream onto two phases of a

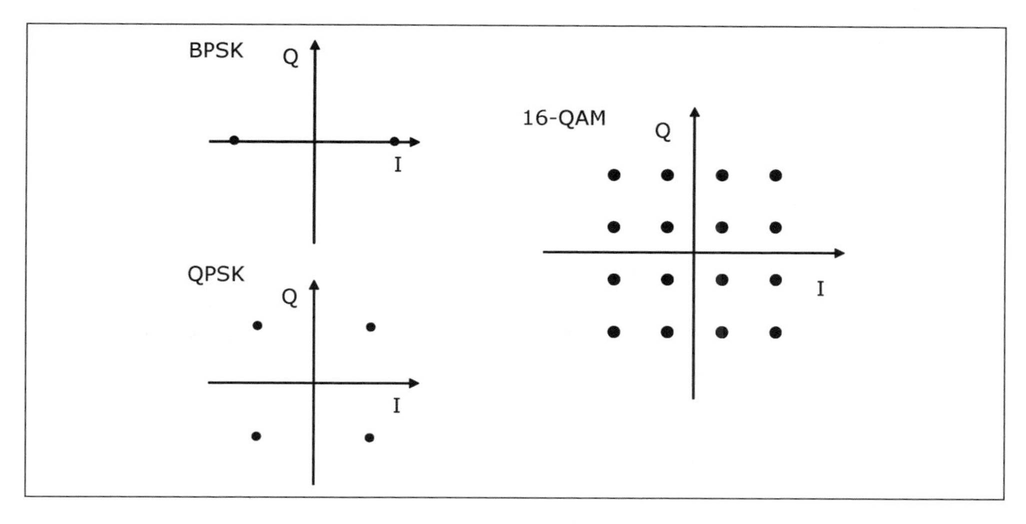

Figure 2.7 BPSK, QPSK, and 16-QAM modulation patterns in the in-phase (I) and quadrature (Q) constellation

signal constellation as shown in Figure 2.7. Each signal point, referred to as a *symbol*, can carry one data bit. In a 4-PSK system, or (quadrature) QPSK, four (2^2) constellation points exist, and therefore a QPSK symbol can carry two bits of data, twice as many as BPSK. In a 16-QAM system, as also shown in Figure 2.7, 16 (2^4) constellation points exist, and therefore a 16-QAM symbol carries 4 bits of data, further increasing system capacity. Presently, the largest constellation point for wireless systems is a 64-QAM system, where each symbol carries 6 data bits ($64 = 2^6$). The information-carrying capacity of different modulation schemes is summarised in Table 2.1.

PSK systems have an advantage over QAM systems in that all their signals have the same amplitude. In contrast, QAM systems have multi-level constellation symbols. This is significant in the design of the transmitter power amplifier as discussed in Section 2.4.

Table 2.1 Modulation schemes and number of information bits per modulated symbol

Modulation scheme	Bits per symbol
BPSK	1
QPSK	2
8-PSK	3
16-QAM	4
64-QAM	6
256-QAM	8

These multi-bit modulation techniques are used to increase the system capacity of a communications system (from fixed-line modems to DSL to broadband wireless). The improvement in capacity is limited by channel conditions as defined by the Shannon Theorem (described in Section 2.8).

2.3 Equalisation

As shown in Figure 2.4, it is important to remove the effects of fading in the channel. We refer to this process broadly as equalisation. Equalisation in a flat-fading channel is equivalent to power control (illustrated in Figure 2.8). By varying the transmission power, the received signal level can be equalised, resulting in a constant received power level. Regardless of the flat or frequency-selective character of channel fading, power control is an important aspect of all systems, and is usually required for channel equalisation. Power control alone, however, is not sufficient at all times and needs to be supplemented by other techniques, which will be discussed in the following text.

In general, channel equalisation is an integral part of both fixed communications and wireless communications. We define three modes of equalisation in broadband wireless systems:

1. Time domain equalisation
2. Frequency domain equalisation
3. Code/multi-user domain equalisation.

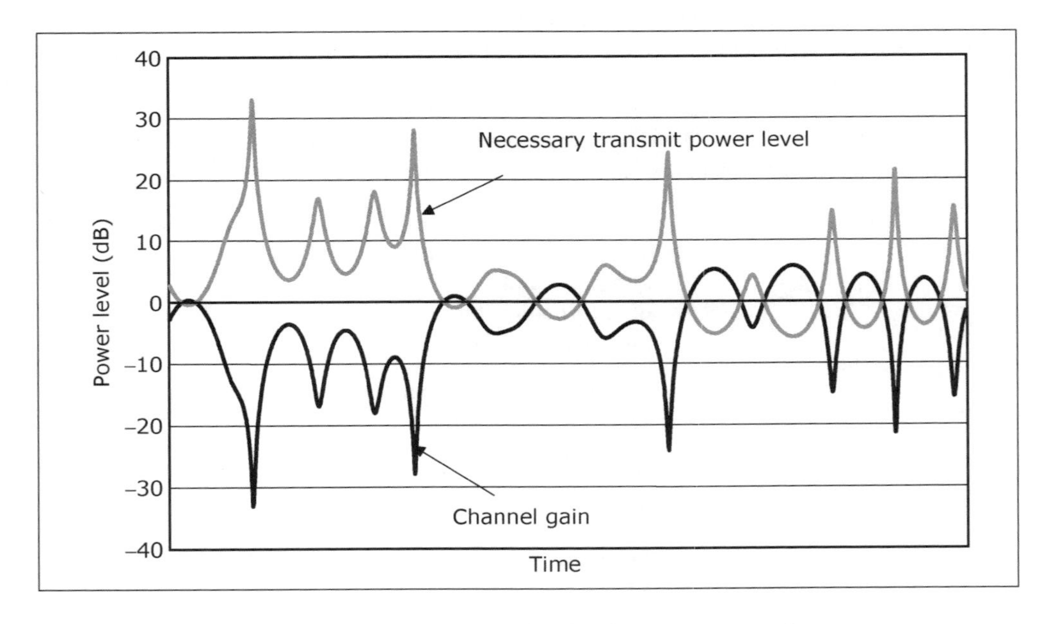

Figure 2.8 Fading equalisation by power control

2.3.1 Time domain equalisation

We define time domain equalisation broadly as power control, a process primarily intended to ensure that a certain received signal level is maintained. The received power level is equal to the transmitted power level multiplied by the channel attenuation. In wireless communications, the attenuation is caused because of three different processes: the distance between the transmitter and the receiver; shadowing due to physical objects blocking the transmission path; and the above-defined fading process. For power control, the transmitter tracks the variation in the received signal level via feedback from the receiver. It then sets its transmission power in order that a certain receiver power level is maintained.

Most power control processes in 2G systems compensated only for distance and shadowing attenuation. This was done in order to maintain a certain performance quality (e.g. BER) and to reduce the level of interference to neighbouring cells, which reuse the same frequencies. Only in IS-95 CDMA systems was power control implemented to equalise fading in the uplink. CDMA-based systems benefit particularly from power control as their frequency reuse factor is one. Moreover, all users transmit in the same band and all of them appear as interference to each other. Therefore, the higher the transmitting power, the greater the interference becomes. Through the power control process, co-channel interference can be reduced. A reduction in co-channel interference leads to an increase in system user capacity.

2.3.2 Frequency domain equalisation

As frequency and time domain signals are closely related to each other, the equalisation processes referred to above can similarly be applied in the frequency domain (see Box: Frequency and time domain conversion).

However, because time domain equalisation involves simple multiplication and addition calculations, it is easier to describe that process. Still, it can be noted that power control can also be carried out by processing in the frequency domain.

Theoretically, equalisation can be done with processing in the frequency domain. A practical realisation of this equalisation can be achieved with a filter that reverses the effect of the transmission channel. For a flat-fading channel, the filter is a simple multiplier, or power control (multiplication by a constant in either time or frequency domains are equivalent). In selective-fading channels, however, frequency equalisation is not generally practical. As can be observed from Figure 2.5, a complex type of filtering would be required

Frequency and time domain conversion
A signal expressed in the time domain can be equivalently and uniquely expressed in the frequency domain. One method to convert a time-domain representation to a frequency-domain representation is through a process called the Fourier transform.The reader is referred to some signal processing texts that define this process in detail; see Further Reading list at the end of this chapter.

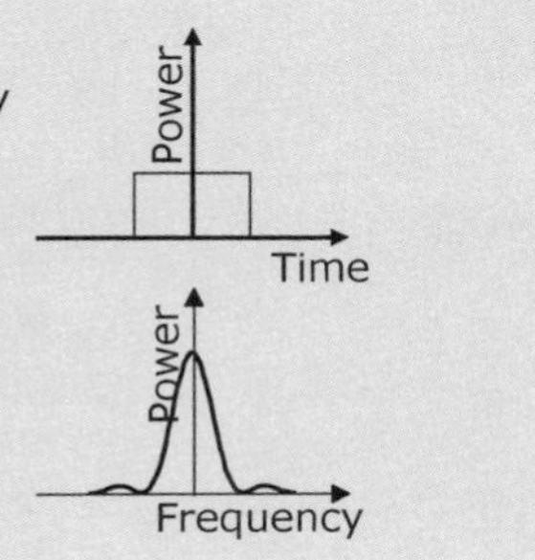

and, in addition, the characteristics of the filter would need to be continuously changed to meet the ongoing channel variations. For these reasons, channel equalisation is performed in the time domain.

Frequency equalisers are often used with fixed-line systems since the channels are more or less stationary. In such an environment, both the receiver and the transmitter are equipped with adaptive filters, which compensate for channel variations. These variations are usually unrelated to fading, as we have described earlier, and mostly depend upon the copper-wire frequency response characteristics.

2.3.3 Code/multi-user domain equalisation

In general, the performance quality of wireless communication systems are limited by two factors. One is receiver noise. Performance depends on whether enough power can be supplied to maintain a certain required signal-to-noise ratio (SNR). The other is co-channel multi-user interference. Here, the performance depends upon the number of co-channel users, the amount of mutual interference they generate, and whether the resulting signal-to-interference ratio (SIR) can be maintained above a required level.

In all multi-user wireless communication systems, user capacity is limited by the amount of interference generated by the signals of co-channel users. When different frequency carriers are used in nearby cells, no immediate co-channel interference exists. However, when the same carrier is used in a neighbouring cell or, as in the case of most CDMA systems where the same carrier is used in all cells, then co-channel interference does exist. We refer to processes designed to reduce or remove co-channel, multi-user interference as equalisation in the code/multi-user domain. In Chapter 3, we list and discuss these processes in more detail.

In CDMA systems, in particular, it is possible to reduce interference by considering the correlation properties of the spreading codes, noting that each transmitter communicates using a distinct spreading code, by which it is distinguished at the receiver side. The multi-user interference can be zero if these codes are orthogonal to each other and transmission is done in a synchronous manner (see Box: Orthogonal spreading codes). This technique is

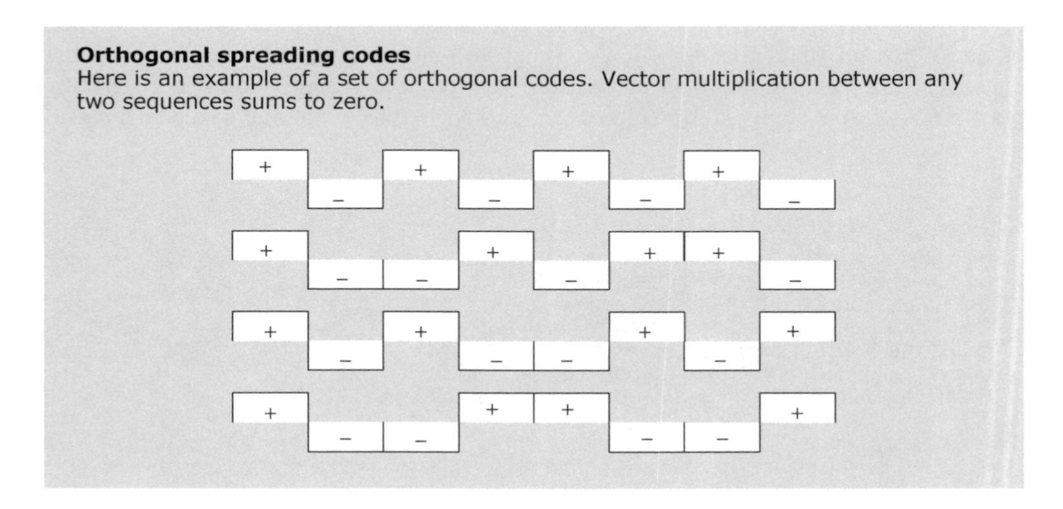
Orthogonal spreading codes
Here is an example of a set of orthogonal codes. Vector multiplication between any two sequences sums to zero.

used in all cdmaOne, WCDMA , and TD-CDMA systems downlink transmissions. However, synchronised orthogonal transmission does not work perfectly under all conditions. Firstly, it is only useful for users in the same cell and in the downlink. Moreover, it is of maximal benefit in flat fading channels: in multi-path channels, the orthogonality is reduced and therefore the multi-user interference is increased.

We define equalisation in the code domain as removing the co-channel user interference caused by sub- or nonorthogonal spreading codes. This may be done through an interference cancellation (IC) or a joint detection (JD) process by restoring and/or creating orthogonality between users' signals through removal of interference. As CDMA systems' multi-user interference is caused by nonzero inter-code correlation values, equalisation is the process of reducing this value to zero through signal processing. We will return to this topic in more detail in Chapter 3.

2.4 Single Carrier and Multi Carrier

Clearly, a communication system's capacity can be increased by increasing the system bandwidth. However, systems with wider bandwidths present opportunities as well as challenges for a communications system engineer. One issue, as we have seen above, is frequency-selective fading. Another issue is how to transmit both low bit-rate signals (associated, for example, with voice communications) and high bit-rate signals (those associated with video communications) over the same wide-band channel.

Two fundamentally different approaches are considered for transmission over a wide-band channel. One is single-carrier transmission, as exemplified in a WCDMA system. Here, the entire bandwidth is used for transmitting a single signal. Low- and high bit-rate signals are then spread over the entire bandwidth using spread-spectrum (SS) technology, which forms the basis of the CDMA systems. The other is multi-carrier technology, where the broadband channel spectrum is divided into many smaller carriers and transmissions are made in parallel over individual narrow-band carriers. Low- and high bit-rate systems may then be accommodated according to the number of carriers they use. The most representative of multi-carrier systems is orthogonal frequency division multiplexing (OFDM), which has been proposed for usage in, for example, WiMAX systems.

2.4.1 Spread spectrum

In one spread-spectrum (SS) technique, a transmitted symbol is multiplied by a signal stream with a much wider bandwidth and, in consequence, the narrow-band signal is spread to encompass the entire bandwidth of the spreading signal. The spreading signal is generally a predetermined code sequence that can be reproduced at the receiver side. Re-multiplication by the spreading code restores the signal to its original narrow bandwidth. This process is illustrated in Figure 2.9.

The ratio of the spread bandwidth to the original bandwidth is called a *spreading factor* (SF) and is usually much greater than one. In 3G CDMA systems, the SF can be as large as 512, but also as low as one (see Box: 3GPP spreading code lengths). Obviously, the lower the SF becomes, the higher the data transmission rate (original signal bandwidth) is.

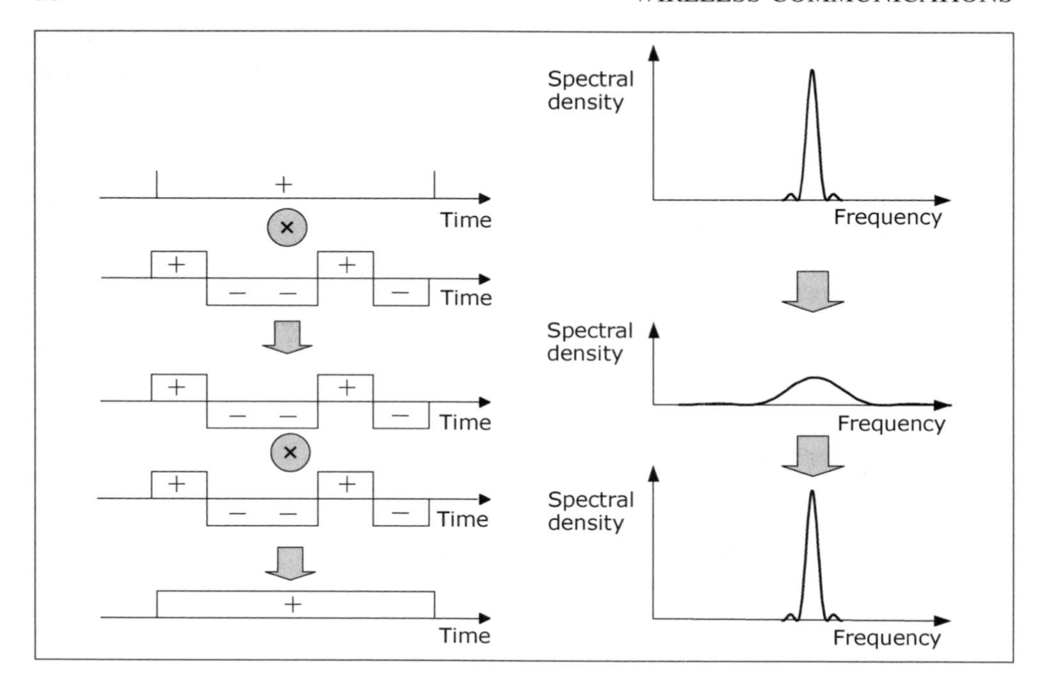

Figure 2.9 Spreading and de-spreading

3GPP spreading code lengths
Two of the 3GPP standards, TD-CDMA and WCDMA, specify the following spreading code lengths:

WCDMA	
Downlink	Uplink
4 ~ 512	4 ~ 256

TD - CDMA	
Downlink	Uplink
1 and 16	1 ~ 16

2.4.2 Orthogonal frequency division multiplexing

In orthogonal frequency division multiplexing (OFDM), the frequency band of operation is subdivided into many smaller, overlapping sub-carriers, as shown in Figure 2.10. Several data symbols are transmitted in parallel over one or more of the sub-carriers. This method of transmission is very efficient over wide-band channels as fading characteristics can be designed to be flat over individual carriers. This is an advantage of multi-carrier systems. Note that if a carrier is sufficiently smaller than the coherence bandwidth, then the fading character is flat. This simplifies the equalisation process described earlier.

Because of the overlapping carriers, transmitter–receiver synchronisation is very important. Furthermore, inter-symbol interference from preceding signals must be avoided.

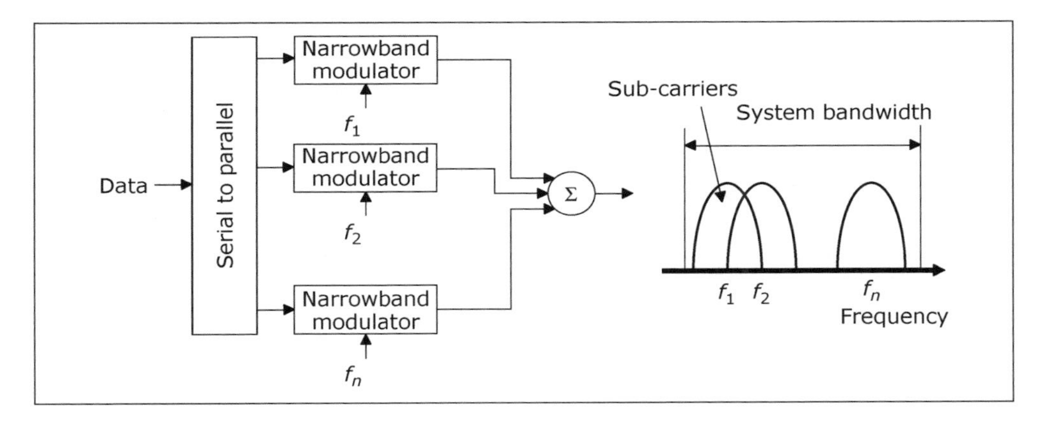

Figure 2.10 Orthogonal frequency division multiplexing

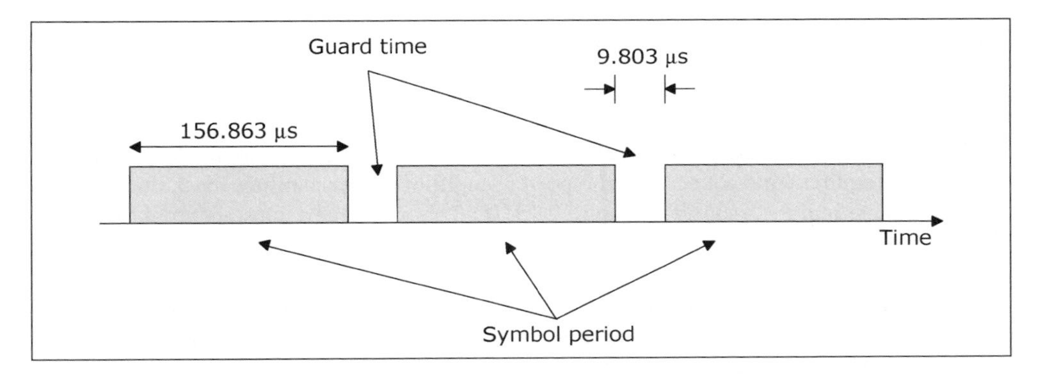

Figure 2.11 OFDM symbol length and guard time in one implementation

Figure 2.11 shows a typical symbol and guard time length. A symbol (e.g. a 16-QAM symbol) is followed by a guard interval whose size is set to be larger than the channel delay spread in order to avoid inter-symbol interference. The symbol is then transmitted over the air using one or more of the OFDM sub-carriers.

2.4.3 Orthogonal frequency-code division multiplexing

Orthogonal frequency-code division multiplexing (OFCDM), as illustrated in Figure 2.12, is a combination of CDMA and OFDM technologies. In OFCDM, the transmitted bit stream is spread in the time domain (as in CDMA systems) or in the frequency domain. As a result, the transmission rate is decreased by the SF. This technology is useful in that it provides an extra spreading gain, (equal to the SF) to a user's received signal. The extra gain can be used to ensure that a signal has adequate SIR: the spreading gain can increase the received power level for users at the cell edge where the signal has been attenuated in comparison with the interference from other co-channel users. As can be seen from Figure 2.12, OFDM and OFCDM processes are very similar.

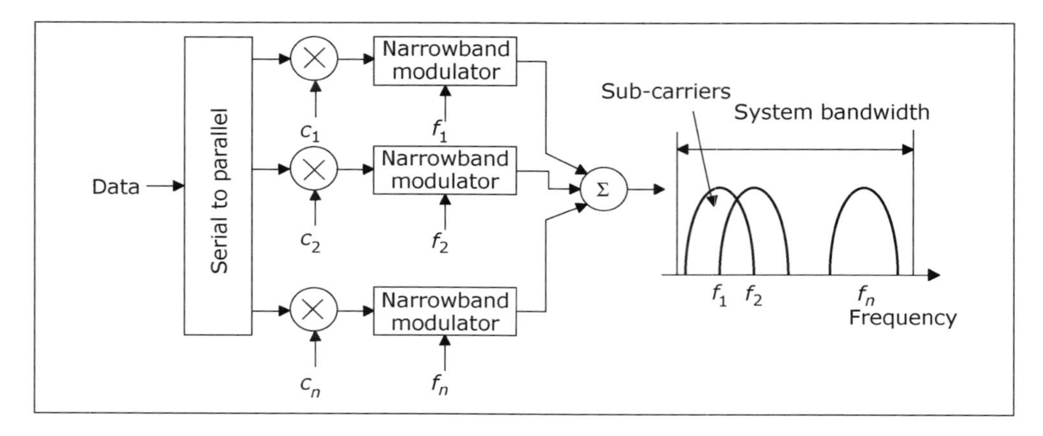

Figure 2.12 Orthogonal frequency-code division multiplexing

2.4.4 Transmission power fluctuation

Both CDMA and OFDM systems suffer from transmission power level fluctuations. This is an important issue as it concerns the transmitter's power amplifier.

In both the systems, use of QAM-based symbol constellation leads to multi-level transmission power requirements. Generally, the power amplifier in a transmitter has a linear region and a nonlinear region as shown in Figure 2.13. If all signals have a constant amplitude, such as in m-PSK signals, then the amplifier may be operated at its maximum amplification level without causing any nonlinearity. This provides a high power utilisation efficiency. However, when signals have multiple amplitude levels, such as in 16-QAM, then the transmitter needs to back off from its highest transmission level, resulting in decreased efficiency.

In OFDM systems, there is another reason for transmission power fluctuation. This is illustrated in Figure 2.14. As an OFDM transmitter combines the signals of several carriers, the combination of these signals in the time domain creates a waveform with a large peak to average power ratio (PAPR). PAPR causes the power amplifier to operate at lower than its optimal operating point. As a result, there is a need for highly linear, voluminous and expensive power amplifiers in OFDM systems. While this may be justifiable in a central base station, it is not practical in hand-held user equipment.

As a result, the practicality of using OFDM-based multiple access systems for uplink transmission is questionable. For the same reason, the usage of modulation of 16-QAM and above in the uplink is not yet implemented in 3GPP. Still, the PAPR issue resulting from 16-QAM modulation is not so significant, and uplink multi-level modulations are being considered by 3GPP. PAPR reduction for OFDM transmission is a major present topic of research.

2.5 Diversity Reception

Power control by itself is incapable of fully compensating for channel fading due to feedback delays and limited transmitter power range. Fortunately, fading mitigation can also be achieved through diversity combining. Several such techniques exist. Two examples are: (1) reception diversity through multiple antennas and (2) multi-path resolution and

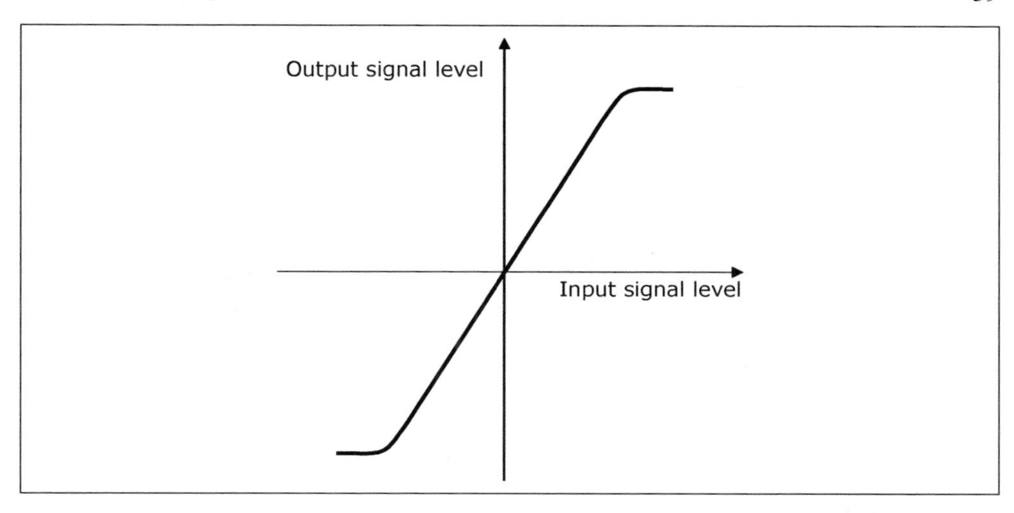

Figure 2.13 Power amplifier characteristics

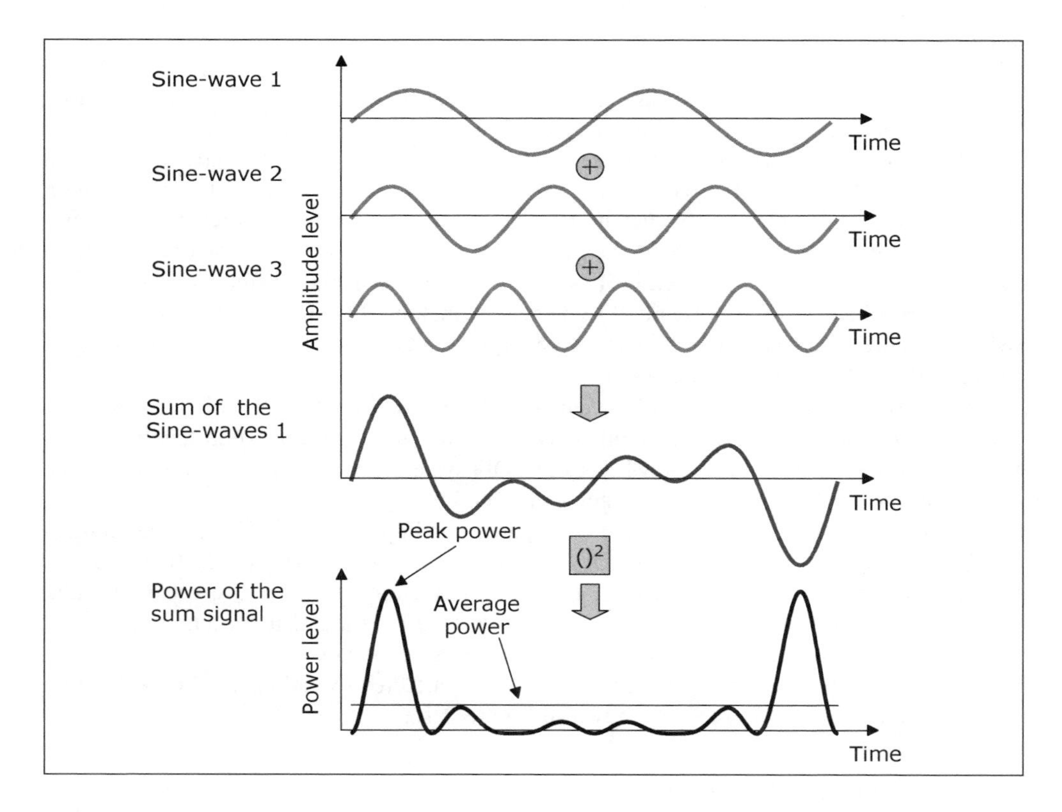

Figure 2.14 OFDM peak to average power ratio

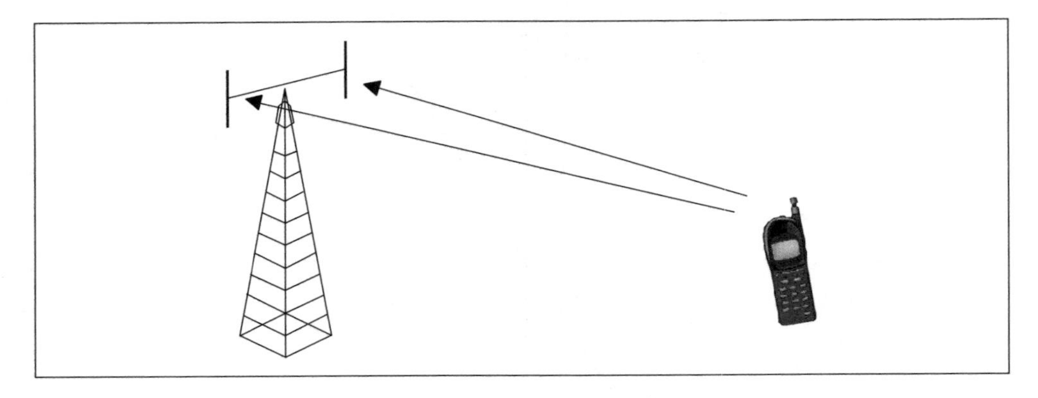

Figure 2.15 Antenna diversity reception

coherent combining. After combining, the fading in one path can be compensated by the received signal from other paths. Since the probability that all paths fade at the same time is significantly less than the probability that a single path fades, the overall received signal experiences less fading.

There are a number of possible methods for achieving independent path diversity. The most common of these methods is antenna diversity, where multiple antennas are used to transmit and receive. This is illustrated in Figure 2.15, where the base station receives signals from a mobile via two antennas. The antennas need to be sufficiently separated in order for signal fading at the receiver antennas to be uncorrelated. Generally, a separation of half a bandwidth ($\lambda/2$) is sufficient for this to occur. This is significant as at times the receiver is not large enough to accommodate two antennas with sufficient separation for reception diversity to be useful. However, as higher and higher frequency bands are used for broadband wireless systems, $\lambda/2$ decreases, making the implementation of diversity reception more practical. Table 2.2 shows the required antenna element separation for some broadband wireless bands in operation.

It is also possible to use multiple antennas at the transmitter side. This is referred to as *transmission diversity*. Transmission diversity is dependent upon information feedback from the receiver side in FDD systems, whereas in TDD systems, channel reciprocity enables an open-loop (no feedback) transmitting antenna selection.

Another diversity technique, which is of particular importance to CDMA communications, is multi-path diversity reception. As discussed above, in frequency-selective channels where the propagation delay spread is larger that the reciprocal of the receiver bandwidth, the signals of any two paths with a propagation delay difference of more than chip period can be separated and independently detected. Since these signals have independent fading patterns, the combination of their signals will be less affected by fading. A simple diagram of a multi-path diversity system is shown in Figure 2.16.

2.5.1 Diversity combining methods

Several diversity combining methods exist for selecting and adding more credible signals, which are received from diverse paths. The combiner usually estimates the signal strength

Table 2.2 Required antenna element separation for diversity reception

Frequency of operation	$\lambda/2$
450 MHz	33.3 cm
800 MHz	18.8 cm
2.0 GHz	7.5 cm
2.5 GHz	6.0 cm
3.4 GHz	4.4 cm
5.3 GHz	2.8 cm
5.8 GHz	2.6 cm

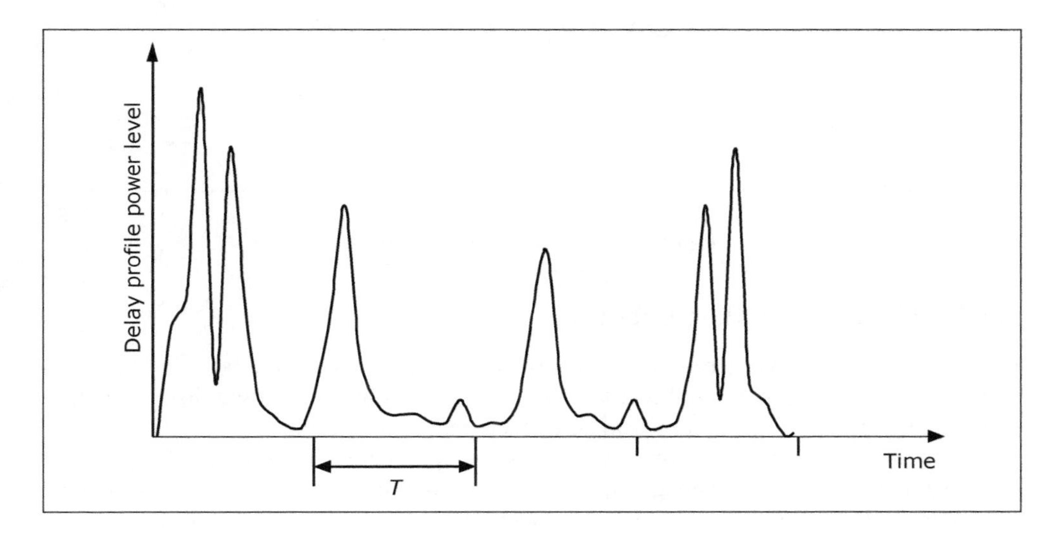

Figure 2.16 Multi-path channel

of each path and sets its combining factors based on these estimates. In Figure 2.16, four independent paths carry the information transmitted from a transmitter to a receiver. A diversity combiner is shown in Figure 2.17. The parameters are set according to the diversity combining method. We define two diversity combining methods: (1) selection diversity and (2) maximal ration diversity.

2.5.2 Selection combining

In this method, the signal from the path with the highest power is selected and the remaining signals are discarded. For example, if path one has the highest received power, the weighing factor for path one is set to a constant and all the other weighting factors are set to zero.

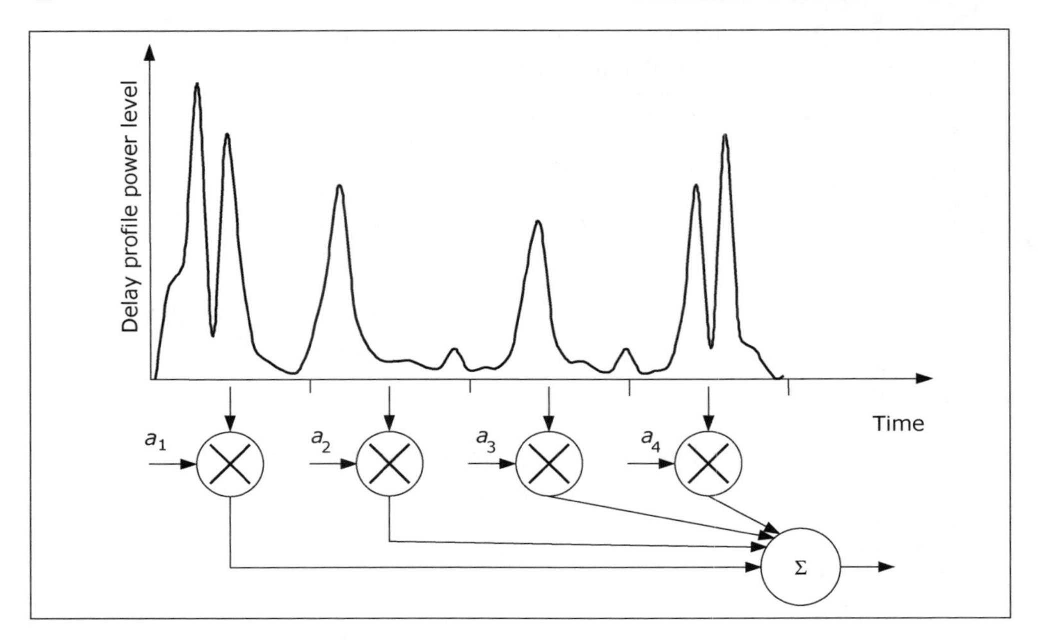

Figure 2.17 A diversity combiner

2.5.3 Maximum ratio combining

In this method, the signals from each path are added together in such a way that the more powerful signals are emphasised and the less reliable ones are suppressed. The weighting factors are now set to values proportional to each path's SNR value.

The performance of a selection diversity combining system is illustrated in Figure 2.18. Note that maximal ratio combining systems perform better than selection combining systems since they add all of the signal energy that enters the receiver. However, both the systems do significantly improve performance as compared to that of flat fading. Note that the performance improvement as order of diversity increases.

2.6 Channel Coding

Another important process in wireless communications is channel coding, also known as forward error correction (FEC) coding. Bits transmitted over a wireless channel are corrupted by the fading and noise described earlier. In order to enable the receiver to recover the transmitted bits from the received information, the transmitted sequence goes through a mathematical process of FEC coding. FEC adds extra information to the transmitted bit stream, which can be used to correct some or all of the received errors. Many FEC schemes have been developed and are in operation. Two of the most recent coding schemes are the most promising candidates for broadband wireless systems. These are (1) turbo codes and (2) low-density parity check (LDPC) codes.

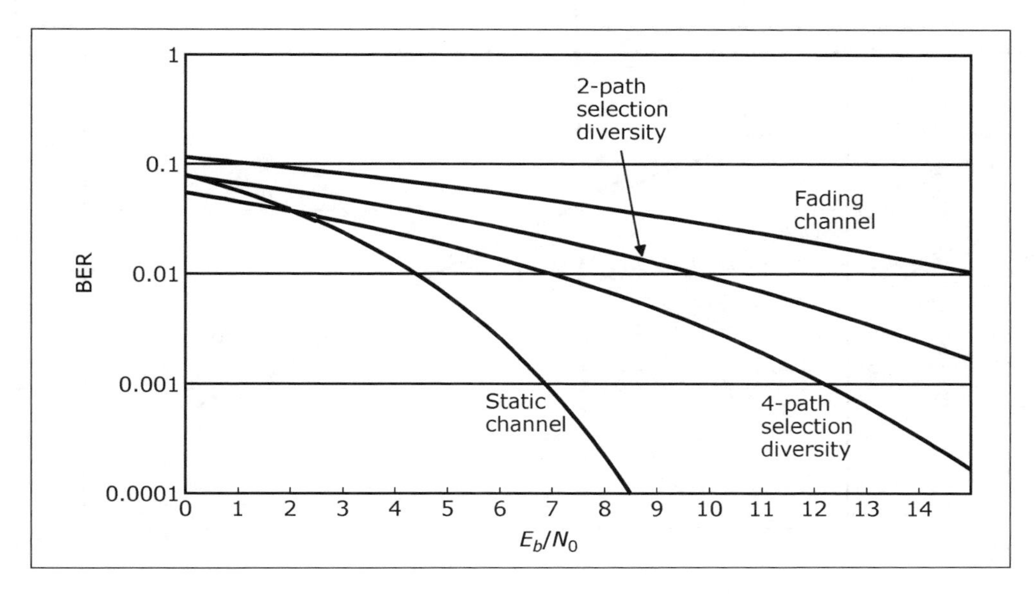

Figure 2.18 BER for diversity combined methods

2.6.1 Turbo codes

First presented in 1993 by Berrou, Glavieux, and Thitimajshima, Turbo codes have been quickly established as the prominent coding scheme for data transmission applications, and are already a part of 3G standards. Their iterative character yields high performance under fading conditions.

2.6.2 LDPC codes

Competitors to Turbo codes are the low-density parity check (LDPC) codes, from the block codes family. This class of codes was invented in 1963 by Gallager. However, their processing complexity was too high for the devices of the day and, therefore, were not implemented in any actual system. With faster processing possible, there has been a renewed interest in these codes over the past few years, and some systems are already using them. LDPC codes are also iterative and perform well under fading conditions associated with radio channels. LDPC codes perform slightly better than turbo codes, and with less processing requirement. The performance of the above two FEC coding techniques in a fading environment is shown in Figure 2.19.

2.6.3 Coding rate

As discussed above, all FEC coding schemes add extra bits to a transmission stream in order to facilitate the correction of errors that may occur during transmission. The ratio

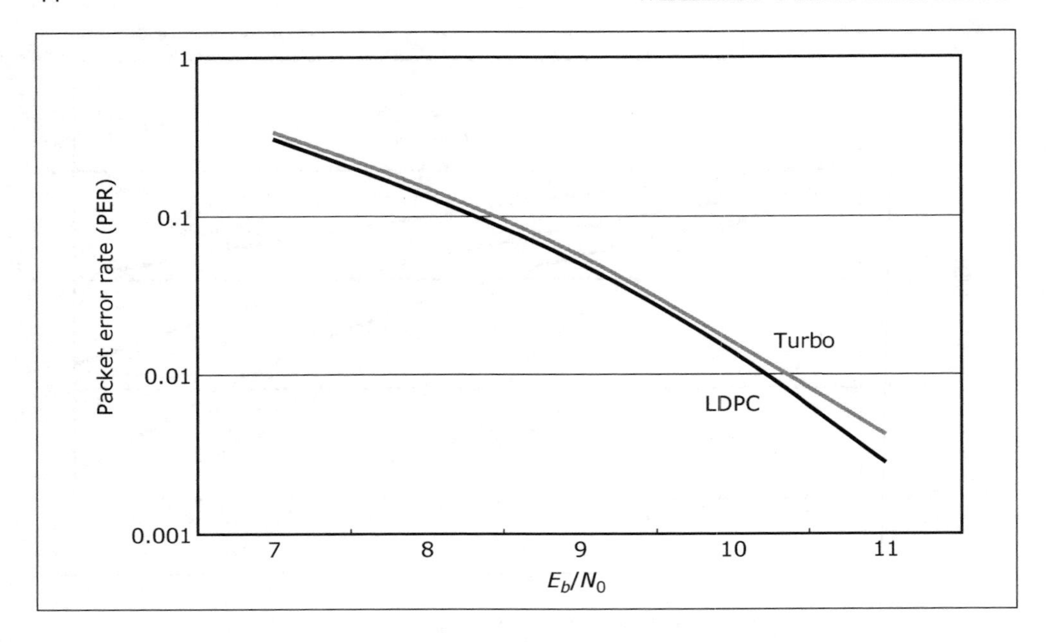

Figure 2.19 BER performance for turbo and LDPC codes

Table 2.3 Coding rates and required SNR

Coding class	Coding rate	Required E_b/N_0
Turbo	1/2	6.6 dB
Turbo	2/3	8.8 dB
Turbo	3/4	10.3 dB
Turbo	8/9	15.2 dB
LDPC	1/2	6.6 dB
LDPC	2/3	8.8 dB
LDPC	3/4	10.2 dB
LDPC	8/9	15.0 dB

of information bits to the total number of transmitted bits is a number between zero and one, and is known as 'coding rate'. The higher the coding rate, the greater the transmission efficiency. However, within an FEC code class, the lower the coding rate, the higher is the error correction capability and the better is the code performance. Table 2.3 shows different coding rates and their required signal quality in terms of signal-to-noise ratio to deliver a 10^{-3} BER in a typical frequency selective fading environment.

2.7 From Circuit Switched to Packet Switched

The transition from a voice-centric circuit-switched system, as is envisaged with 3GPP standards (see Table 1.5), to a data-centric packet-switched system is not trivial. It requires a rethinking of how links are set up between a base station and several types of end-user equipment, and how these resources are to be shared.

In a typical wireless circuit-switched call, a wireless resource is given to the connection between a mobile unit and a base station. The link is dedicated to the call: there are call set-up, call maintenance, and call termination processes. Power control is maintained to ensure maintenance of required signal QoS. Call handover to another base station often occurs in the middle of a call. There are many similarities between a fixed circuit-switched call and a wireless one, in that the base station acts as the last node for delivering a call to an end-user phone.

In comparison, a packet-switched call consists of a stream of many packets, each of which independently travel through the network to a destination. They pass through many switches (or routers) along the way. At each switching point, the packet queues for processing by the router, which looks at a packet's control information and sends the packet onwards to its destination via other routers. This process is illustrated in Figure 2.20. The links between routers are shared resources and used temporarily by each packet. A 'permanent' link between two end-users does not exist.

What does this process mean for wireless packet switching? How do routing and resource sharing translate into wireless communications? And how are call maintenance functions, such as power control, carried out in a wireless packet call?

2.7.1 Shared channels

A packet-switched wireless communications system is illustrated in Figure 2.21. Here, the base station acts as the last router on the way to the end-user in the downlink. In the uplink, a base station is the first router, and switches the packet onwards to its destination via other routers. The base station in effect acts as a server: packets arrive for downlink transmissions, queue at the base station's buffer, and are served in turn. In the uplink, user

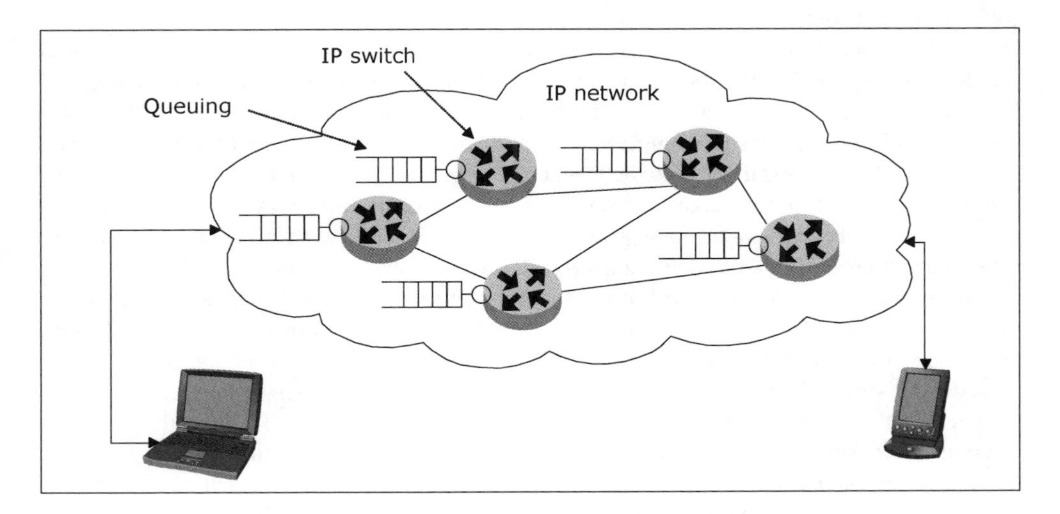

Figure 2.20 Fixed-line packet switching

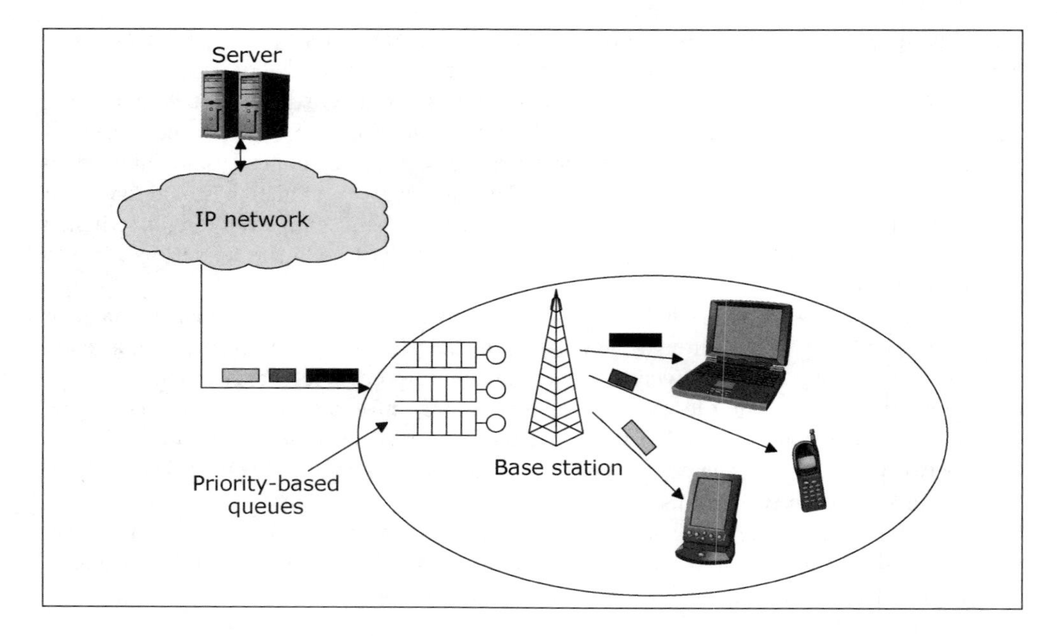

Figure 2.21 Downlink wireless packet queuing and switching

equipment sends transmission requests to the base station, these requests are received and queued, and transmission requests are granted in turn.

The server may serve down- and uplink transmissions on a one-by-one basis, similar to a single-server queue. Here, a base station transmits to only one user and receives from only one user at a time. It may also be a multiple-server queue, and transmit and receive to/from several users.

While in fixed-line systems such a single/multiple server is simple to implement, a wireless server needs to consider many extra aspects. Among these are power control, handover, and resource allocation. Furthermore, downlink and uplink connections base station and end-user equipment cannot be dedicated, and need to be shared with other users. Care must also be taken so that the packets do not collide. In the downlink, the base station can control and arrange all transmissions in a centralised way so that packets do not collide. In the uplink, the base station can maintain collision-free communication by scheduling transmissions of each connecting UE over a shared channel.

In particular, the power control process is difficult to implement. In the downlink, power control is not critical, as adaptive coding and modulation can take advantage of good channel conditions. In the uplink, however, a feedback process needs to be established for power control. Individual feedback channels are difficult to establish, which is one reason shared channels in the uplink of FDD systems have not yet been implemented. Currently, there are ongoing discussions in implementing an uplink shared channel in future releases of 3GPP WCDMA. In TDD systems such as TD-CDMA, however, power control can be carried out in an open-loop fashion. This has facilitated the specification and implementation of uplink shared channels.

2.7.2 Packet scheduling

In fixed-line packet switching, packets are queued as they arrive at a server. They are however, processed according to the priorities each packet has. For example, a VoIP packet gets priority over a file-transfer packet as it may be more delay sensitive. A base station server must also consider this. In addition, as users can transmit and receive at different rates according to their locations within the cell, the server must decide which user may be serviced first so that the total system throughput is maximised.

Independent of scheduling priority based on the type of traffic, three main scheduling algorithms have been defined for downlink and uplink packet transmissions. These are (1) round robin, (2) maximal SINR, and (3) fair proportional.

Round robin

In this scheduling algorithm, the base station transmits to and receives from users within a cell without any consideration of their relative location. End-user traffic is basically served on a first-come, first served basis. This algorithm ensures fairness, but system throughput is less than optimal.

Maximal SINR

In this scheduling algorithm, the base station transmits to and receives from end-user equipment in decreasing order of their SINR. This method delivers the highest possible throughput for the system in total. However, it is not a fair algorithm in that it tends to allocate resources unfairly to users nearer the base station at the expense of users farther away.

Fair proportional

This algorithm combines the previous two algorithms. Its performance is somewhere between the two: a fairer resource allocation, at a somewhat less than maximum total throughput.

2.7.3 Header compression

The link between a base station and an end-user equipment is the outermost link in a call connection. In contrast to a fixed, wireline connection, most of the information contained in the IP packet header are not necessary for each individual packet. This is because most of the IP packet header information, such as the type of service, source, destination, and so on, are repetitive, and remain fixed for one entire call. These need to be transmitted only once. Packet header compression techniques are designed to remove these redundancies, and can reduce the IP packet header size by more than 90% on average.

2.7.4 Wireless VoIP

How is voice carried effectively over an all-IP network? VoIP in fixed systems works much like the packet-switched process defined earlier. Some priority queuing may be implemented as VoIP packets are delay sensitive.

Voice over IP

VoIP from PC services have been available for some time. Examples are Yahoo Messenger and MSN Messenger. However, VoIP from PCs really took off with Skype, whose high voice and service quality have attracted many users.To date, there have been more than 147 million downloads of the program. Skype Out, a service introduced in 2004, allows users to call an ordinary phone number from a PC at very low rates.
Many VoIP operators from normal phones have also emerged. In Japan,Softbank's BB-phone, a service associated with YahooBB ADSL service is a good example.

Table 2.4 QoS classes

Parameters	Conversational	Streaming	Interactive	Background
Maximum bit rate (kbps)	<2048	<2048	<2048	<2048
Maximum packet size (Bytes)	$\leq$1502	$\leq$1502	$\leq$1502	$\leq$1502
Packet error rate	$10^{-2} \sim 10^{-5}$	$10^{-1} \sim 10^{-5}$	$10^{-3} \sim 10^{-6}$	$10^{-3} \sim 10^{-6}$
Delay (ms)	100	250	Not defined	Not defined

However, as voice traffic is in effect a continuous process, and the fact that usually it is highly delay sensitive, base station call set-up and scheduling algorithms prioritise this type of calls. This is done by allocating transmission resources in such a way as to ensure a voice call is not cut off in the middle of a conversation and that transmission delay is maintained below a design threshold (see Box: Voice over IP).

2.7.5 Quality of service

Different classes of services have been defined in 3GPP standards, and for each service, the delay and acceptable error probability levels have been specified. These are shown in Table 2.4. Four different classes of services have been defined in 3GPP. Conversational refers to voice and video conferencing type of services where strict delay constraints exist. Streaming services have tight delay and error rate constraints. Interactive services, such as text messaging, have strict error rate constraints, while delay is not defined. Background refers to services such as file transfer, which have strict error rate requirements, but no delay constraints.

2.8 System Capacity

Capacity is generally defined as the amount of information that can be transmitted with a designed level of QoS over a certain bandwidth. Capacity is a broad term and denotes the maximum ability of a medium to transfer information. A related term is frequency-utilisation efficiency, and refers to the actual data bits transferred over a certain bandwidth of a mobile communications system. In 1G and 2G services, the efficiency generally referred to the number of voice channels that could be simultaneously supported. In data communication services, this definition has evolved to mean the total number of information bits that are correctly transmitted per second over a service area.

The capacity of a system is also evaluated in relation to the amount of offered traffic and the ability of the system to process them with small call rejection (call blocking) and in a timely manner. Trunking efficiency refers to the system capacity (number of voice channels, or throughput capacity) required to process a certain amount of offered traffic in terms of voice calls or data packets with minimum blocking or delay.

2.8.1 Shannon theorem

In general, the system capacity, or absolute maximum usage efficiency, can be calculated by the Shannon Theorem. Claude E. Shannon, known as the father of information theory – and, by extension, considered by many as the father of modern wireless communications – defined the capacity limit of a communications system, C, in information bits per second, as depending on the ratio of signal power, S, to noise power, N, as below:

$$C = B \log_2 \left(1 + \frac{S}{N}\right)$$

where B (in Hz) represents the signal bandwidth. The maximum usage efficiency η (in bits/second/Hz) is therefore capacity divided by bandwidth:

$$\eta = \frac{C}{B} = \log_2 \left(1 + \frac{S}{N}\right)$$

These equations show that the capacity of a communications system is proportional to the available bandwidth. It also increases as the signal quality in terms of the SNR ($\frac{S}{N}$) improves. To increase system throughput, higher order modulations, such as those defined in Section 2.2.1, for example, 16-QAM, should be used. However, to deliver these higher capacities, a higher SNR needs to be maintained . The equation also shows that frequency utilisation efficiency is dependent only on SNR, and to the degree that consistent signal quality can be maintained throughout an operational area. We will discuss more about capacity in Section 3.2.

2.8.2 Trunking efficiency

Traffic arrival is a random process. For example, in circuit-switched voice communications, while subscribers served by a single base station may generate traffic in such a way that on average 20 voice channels are busy, at times more and at times fewer calls are in process. In order to cope with an average of 20 calls, more than 20 channels are required to ensure that none, or very few, are blocked. The same applies to data communications. The difference is that while because of their real-time nature voice calls must not be blocked, data packets may be queued at a buffer and transmitted later. Here, the operator must ensure that delay due to queuing remains below a set level. The probability of blocking and the probability of queuing are known also as grade of service (GoS). GoS for voice is usually required to be 1%. For packet communications, acceptable delay caused by queuing varies according to each service, as shown in Table 2.4. This field of telecommunications is known as *traffic engineering*.

In fixed systems, where the number of channels (circuits) is well known and fixed, traffic engineering models and units are well researched. The amount of offered traffic (or traffic generated by users within a cell) is measured in units of Erlang. One Erlang traffic is the traffic that would continuously occupy one channel on average. Because of the random nature of traffic as discussed earlier, however, more than one channel is required to carry one Erlang of traffic with low blocking probability. As the amount of traffic increases, the variation in total, or difference between peak and trough, decreases with respect to the total. Therefore, the amount of traffic that can be carried with an acceptable GoS approaches the number of circuits. This is known a trunking efficiency, where the more the number of channels (trunks), the more efficiently they will be utilised. The probability of blocking (for voice) is calculated from 'Erlang-B' formula, and is shown in Table 2.5. It can be seen that, for example, trunking efficiency increases from 1% in one channel to 44% in 10 channels, and to 84% and 90% for 100 and 200 channels.

A separate formula, "Erlang-C", is used for traffic streams that can be queued, such as those associated with packet-switched data communications. Erlang-C formula calculates the probability of queuing and the average delay for ***constant-sized*** packet traffic, and a special class of traffic arrival (known as exponentially distributed inter-arrival). As such, it is not applicable to the traffic experienced in mobile communications, where different classes of packet traffic coexist, and the traffic behaviour of each stream is by and large independent. For such a traffic system, the calculation of queuing probability is complex and depends on the packet size and the number of different data stream types. A sample of queuing probability and average delay calculations are presented in Table 2.6 as a general indication of delay behaviour. It should be noted that in many instances it is the delay variations that are of interest, for example, what the maximum delay is and what percentage

Table 2.5 Possible traffic throughput for a 1% grade of service

Number of channels	Possible traffic (Erlangs)
1	0.01
2	0.15
3	0.46
5	1.36
7	2.50
10	4.46
20	12.03
30	20.34
50	37.90
70	56.11
100	84.07
200	179.74

Table 2.6 Queuing probability and average delay figures for different amounts of traffic/circuit as calculated from the Erlang-C formula (average packet call length = 10 s)

Number of channels	Offered traffic (Erlangs)	Queuing probability	Average delay
20	10	0.4%	0.0
20	16	25%	0.6
20	19	75%	7.5
40	20	0%	0.0
40	32	12%	0.2
40	38	66%	3.3
80	40	0%	0.0
80	64	4%	0.0
80	76	54%	1.4

of users experience a delay larger than a certain threshold. Erlang-C formula is nowadays mostly used to estimate the number of required operators at call centres. On the basis of estimates of the average number of callers at any time, it calculates the number of necessary operators so that the callers do not experience long holding times.

Furthermore, in wireless systems, it is difficult to determine the number of circuits with high accuracy, and therefore traffic engineering is only used as a general guideline. Nevertheless, the concept of trunking efficiency itself for both the data voice and data traffic is equally applicable to fixed and wireless communications.

2.9 Coverage

Electromagnetic waves that carry radio signals over a transmission channel lose their power as they travel further from the transmitter. In a mobile communications environment, they also lose power as they pass through, or reflect off, physical objects such as buildings, and so on. A received signal power needs to be received with a certain minimum level as compared to locally generated noise power (as well as other interference) for it to be detectable by the radio receiver. In the downlink, signal level drops the further it travels from the base station transmitter, and the area within which signal power remains above the required minimum is known as 'coverage area'.

The maximum area covered by a BS is a function of the BS's transmission power and the frequency of operation. One model often used for calculating the received power level in mobile communications environment is the extended Hata model. In this model, the average path loss L (in dB), as the ratio of the received power to the transmitter power, is a function of carrier frequency, f_c (in MHz), and distance, d (in km), and transmitter and receiver antenna heights. A simplified equation for the urban environment, assuming antenna heights of 30 m for the transmitter and 1.5 m for the receiver is as follows:

$$L \simeq 26 + 34\log(f_c) + 10\log\left(\frac{f_c}{2000}\right) + 35\log(d)$$

This equation is valid for an f_c of 2 to 3 GHz, covering a major part of the frequencies of interest in mobile communications. This equation shows that path loss is related to distance raised to the power of 3.5 and to carrier frequency raised to the power of 3.4. Frequency bands considered for broadband wireless are listed in Figure 1.7. As discussed in Section 2.5, the realisation of diversity reception is easier at higher carrier frequencies because of relatively smaller inter-antenna separation (Table 2.2). However, studying the above equation, we find there is also a disadvantage in operating at higher frequency bands: propagation loss increases as the frequency band of operation increases. The following equation illustrates the ratio of path loss L_R in going from a carrier frequency of f_1 to f_2.

$$L_R \simeq \left(\frac{f_2}{f_1}\right)^{3.4}$$

Operating, for example, in the 3 GHz band instead of the 2 GHz band results in a quadrupling of the path loss. In other words, the received power is four times lower on average. It is therefore desirable to operate in lower frequency bands.

Coupling this with the extra power required for higher transmission rates, the area covered by a base station can diminish greatly. One solution to this problem is to increase the power of the transmitters. However, this soon becomes impractical because of amplifier costs and power consumption constraints. Another way is to install more base stations, which again means higher infrastructure costs. New network topologies have also been offered as a solution. We will discuss these in Chapter 4.

2.9.1 Link budget

In order to determine the required transmission power and the appropriate type of antenna equipment for both the base stations and end-user equipment, a technique called *link budget* is utilised. Link budget is an old technique used in all communication systems, from wireline telephony to deep-space satellite communications. The premise is that by considering the required receiver signal power to noise and interference ratio, signal attenuation per distance, receiver noise and interference level, and transmission power and antenna gains, one can calculate the maximum distance a base station can cover. Table 2.7 shows an example of a typical link budget calculation for a wireless communications system.

2.9.2 Multi-hop

Multi-hop communication refers to a technique where downlink transmissions from a base station to far-away end-user equipment (UEs) are carried via intermediate equipments. The same applies to uplink transmissions. The multi-hop concept is illustrated in Figure 2.22. Multi-hop communications introduces a new layer of complexity in all the issues discussed above, including scheduling, routing, shared channels, and so on. We will discuss this and other new network topologies in Chapter 4.

Table 2.7 An example of a link budget for the downlink section of a wireless communications system with a 2 MHz bandwidth, operating at 2 GHz

Parameters	Value
(a) Transmission power	37 dBm
(b) Base station antenna gain	17 dB
(c) Transmitter cable loss	4 dB
(d) Receiver antenna gain	0 dB
(e) Noise power	−111 dBm
(f) Interference margin	3 dB
(g) Required signal to noise and interference ratio	7 dB
(h) Required received power level = e + f + g	−117 dBm
(i) Fading margin	5 dB
(j) Building penetration loss	6 dB
(k) Maximum allowable signal loss = a + b − c + d − h − i − j	140 dB
Maximum range according to the Hata model	1.1 km

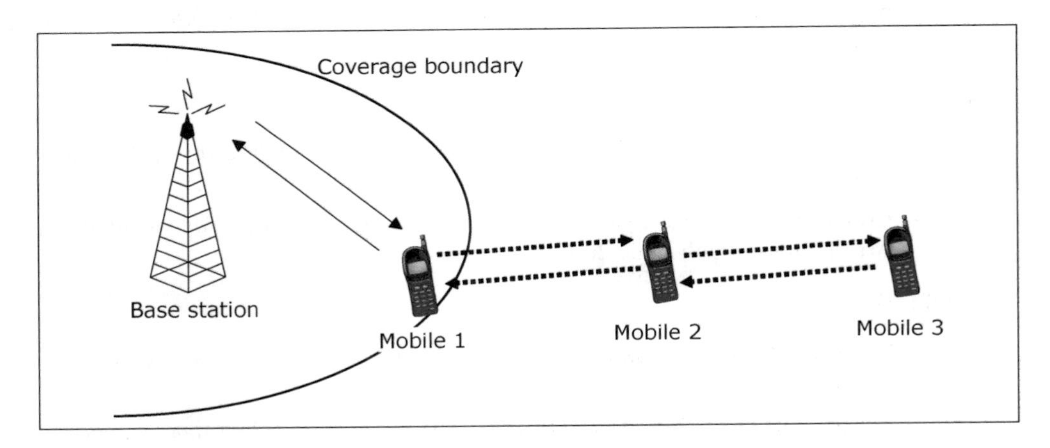

Figure 2.22 Multi-hop communications

Summary

In this chapter, we have given an introduction to the technical issues of broadband wireless communications. We discussed channel fading caused by multi-path signal arrival, and how candidate technologies compensate for this fading. How to equalise for channel conditions is a major starting point for many, if not all, wireless technologies. We then discussed some of the technologies related to wireless transmissions, the transition form circuit switching to packet switching and various QoS issues. Finally, we touched on the subject of coverage and new network topologies for extending the coverage of a base station.

This chapter was intended to provide a summary understanding to the technical issues of broadband wireless and form a reference point for the technical issues that will come up in relation to costs in the succeeding chapters.

Further Reading

- On Wireless and Mobile communications
  - Pahlavan, K., Levesque, A. H., *Wireless Information Networks*, John Wiley & Sons, 1995.
  - Rappaport, T. S., *Wireless Communications: Principles and Practice, 2nd Edition)*, Prentice Hall, 2001.
- On CDMA:
  - Holma, H., Toskala, A. (Editors),*WCDMA for UMTS: Radio Access for Third Generation Mobile Communications, 3rd Edition*, John Wiley & Sons.
  - by Viterbi, A. J., *CDMA : Principles of Spread Spectrum Communication*, Addison-Wesley, 1995.
- TD-CDMA
  - Chitrapu, P., *Wideband TDD : WCDMA for the Unpaired Spectrum*, John Wiley & Sons, 2004.
  - Esmailzadeh, R., Nakagawa, M., *TDD-CDMA for Wireless Communications*, Artech House Publishers, 2002.
- OFDM
  - Fazel, K., Kaiser, S., *Multi-Carrier and Spread Spectrum Systems* John Wiley & Sons, 2003.
  - van Nee, R. D. J., Prasad, R., *OFDM for Wireless Multimedia Communications*, Artech House Publishers, 2000.
- 3GPP standards
  - 3GPP web site: `http://www.3gpp.org/`, 2005.
- Joint detection and Interference cancellation
  - As implemented by IPWireless Inc. Web site:
IPWireless `http://www.ipwireless.com/`, 2005.
  - Verdu, S., *Multiuser Detection*, Cambridge University Press, 1998.

- On Turbo codes
  - Berrou, C., Glavieux, A., Thitimajshima, P., Near Shannon limit error-correcting coding and decoding: Turbo-codes, *Conference Record, IEEE International Conference on Communications, ICC*, 1993, Geneva.
  - Turbo codes book Vucetic, B., Yuan, J., *Turbo Codes: Principles and Applications*, Springer, 2000.
  - Web site with other references and codes: `http://www.csee.wvu.edu/ mvalenti/turbo.html`, 2005.
- LDPC codes
  - David MacKay's Gallager code resources, web site: `http://www.inference.phy.cam.ac.uk/mackay/CodesFiles.html`, 2005.
  - Gallager, R., Low-density parity-check codes, *IRE Transactions on Information Theory*, Vol. 8, PP. 21–28, 1962.
- On Link budget:
  - Text books on satellite communications, e.g. Kadsh, J. E., East, T. W. R., *Satellite Communications Fundamentals*, Artech House Publishers, 2000.
  - General purpose link budget calculator, NICT web site: `http://w3.antd.nist.gov/wctg/manet/prd_linkbudgetcalc.html`, 2005.
- On Traffic engineering
  - Kleinrock, L., *Queuing Systems*, John Wiley & Sons, 1975.
  - Web site for calculating blocking and queuing probabilities: `http://personal.telefonica.terra.es/web/vr/erlang/eng/cerlangc.htm`, 2005.

3

Enhancing Technologies

In Chapter 2, we discussed a number of present technologies that are used for wireless communications. We also defined the capacity of a cellular system and the area a base station can cover. Use of any of the technologies described in Chapter 2 leads to a specific performance in terms of capacity and coverage. Although the maximum capacity is limited by the Shannon theorem for all systems, the way each system uses the frequency resource leads to different frequency usage efficiencies.

In this chapter, we discuss techniques that can further increase the capacity of a particular wireless communications system as well as its maximum coverage. We classify these techniques into two broad groups. In the first group, techniques are dependant upon the communications technology being used, and are generally based on advanced signal processing. The degree of enhancement is also dependant upon the particular technology. Although many of these techniques are applicable to most transmission technologies, some are particularly applicable to a few, and therefore can make a difference in the systems' overall performance. Interference reduction and channel equalisation are among these techniques. The other techniques are communications technology independent: they can generally be applied to any technology and require minimal changes of the technology. Hybrid automatic repeat request (HARQ) and advanced antennas are among these techniques.

For antenna technologies, we will consider sector, adaptive array, and multi-input multi-output (MIMO) antennas. Generally, antennas can be designed to serve both of the above performance enhancement purposes: increasing coverage and/or increasing capacity. As antennas are 'external', they can be applied to almost any of the technologies listed in Chapter 2. The caveat is that some of the transmission technologies cannot fully take advantage of these antennas. We define and evaluate these techniques with reference to system capacity as determined by the Shannon theorem. But first, we further describe performance limitations in terms of noise and interference and how frequency reuse affects performance.

3.1 Frequency Reuse

As discussed in previous chapters, wireless access technologies provide service to a large number of users simultaneously, using a limited bandwidth. Bandwidth resource sharing

Broadband Wireless Communications Business: An Introduction to the Costs and Benefits of New Technologies Riaz Esmailzadeh

between users must consider two issues. First, when two equipments use the same frequency band to transmit or receive, their signals appear as interference to each other. Therefore, to the extent possible, the same frequency should not be used simultaneously by two or more users. However, two users may use the same frequency if, physically, they are sufficiently far apart. This means that interference would remain sufficiently small so as not to detract from the mutual quality of communications. This is the fundamental principle behind cellular communications (and cell phones). In cellular systems, the area of service coverage is divided into cells. Each cell uses a certain portion of the available frequency spectrum. The same portion may be used in another cell so long as the two cells are sufficiently far apart. A parameter called *cell reuse factor* indicates how often the same spectrum is used. This is illustrated in Figure 3.1, where the cell reuse factor is equal to 7 (typical for 1G systems). The access technologies define how several end-user equipment connect to a central base station as shown in Figure 1.1. The access technology characteristics determine how much interference each user generates and can tolerate. This in turn determines the frequency reuse factor, and the frequency-utilisation efficiency.

In general, wireless communication systems' performances are either noise limited or interference limited.

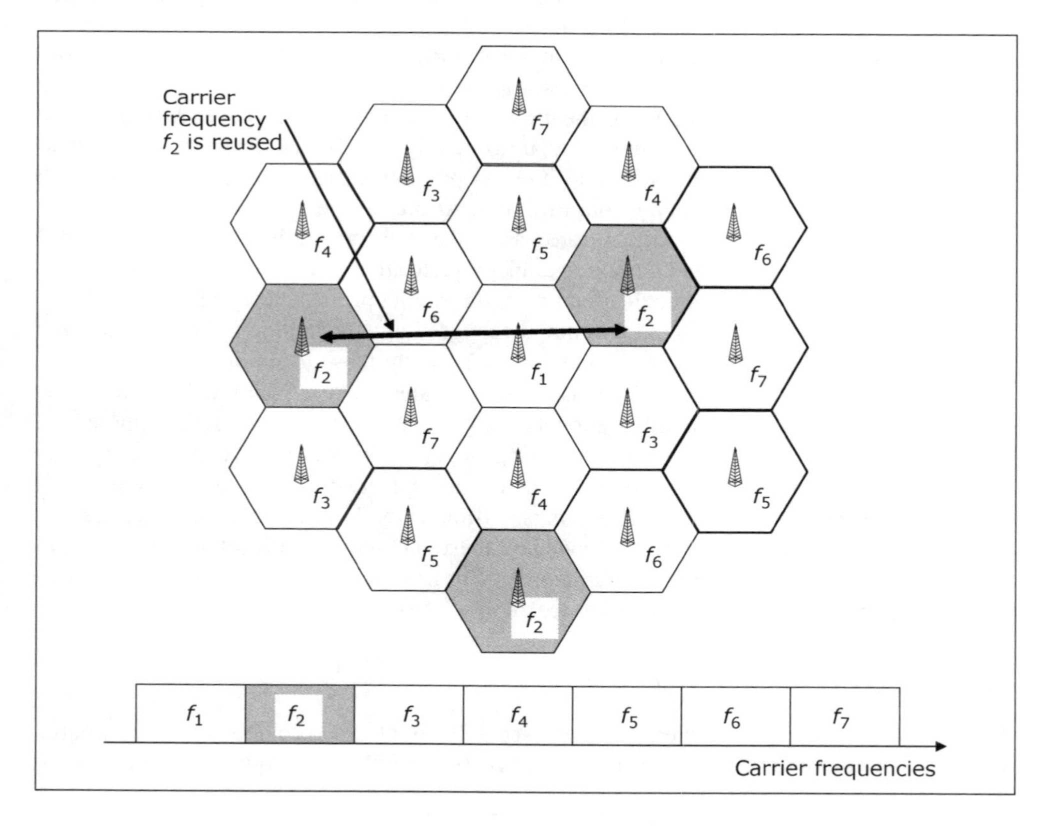

Figure 3.1 Cellular communications and frequency reuse

3.1.1 Noise limited

In noise-limited systems, the required signal-to-noise ratio (SNR) to maintain a certain degree of performance is significantly larger than the signal to co-channel interference ratio (SIR). Through frequency reuse, the level of interference power within a frequency carrier is kept to well below the noise power level. Noise, and in particular thermal noise, which is significant in wireless communications, is produced through a physical process and cannot be avoided. It is a function of receiver bandwidth and operating temperature. Considering that signal power level decreases with distance, approximate distances necessary for tolerable signal decay can be calculated. On the basis of this, the frequency reuse factor is found.

Frequency reuse in noise-limited systems is utilised basically because these systems are not designed to tolerate interference well. For example, all first-generation analogue systems utilised frequency reuse to operate as noise-limited systems. In these systems, interference was of great concern, as co-channel users could actually be heard as cross talk. Second-generation systems such as global system for mobile communications (GSM) and personal digital cellular (PDC) had a much tighter frequency reuse and were, for the most part, noise limited. However, here too, excessive interference cannot be well tolerated.

3.1.2 Interference limited

Interference-limited systems are designed to operate under conditions where co-channel users are present, and thus a designed degree of interference can be tolerated. For these systems, SNR is sufficiently higher than SIR, so SNR does not usually feature in system performance evaluation. CDMA systems are examples of interference-limited systems, where co-channel users are mutual interferers, and can be detected through the de-spreading process.

A degree of performance quality can be maintained so long as interference is kept below a certain level (usually well above noise level). Moreover, these systems process signals in such a way that interferers appear to each other as noncoherent noise-like signals. Cross talk, or intentional eavesdropping, is made difficult through spreading or other scrambling processes. Frequency reuse may still be used in interference-limited systems. However, it is used here in order to increase a particular cell's user throughput, or for particular cellular topologies as will be discussed in Chapter 4. The broadband wireless systems of concern to this book are generally all interference limited, although in a few cases interference is reduced by low frequency reuse factors.

3.2 Capacity Limit

Looking again at the bandwidth-normalised Shannon theorem, we can observe that the ratio of a system's capacity (or throughput) C to its signal bandwidth B, is dependant on the ratio of signal power S to noise power N (SNR):

$$\frac{C}{B} = \log_2\left(1 + \frac{S}{N}\right)$$

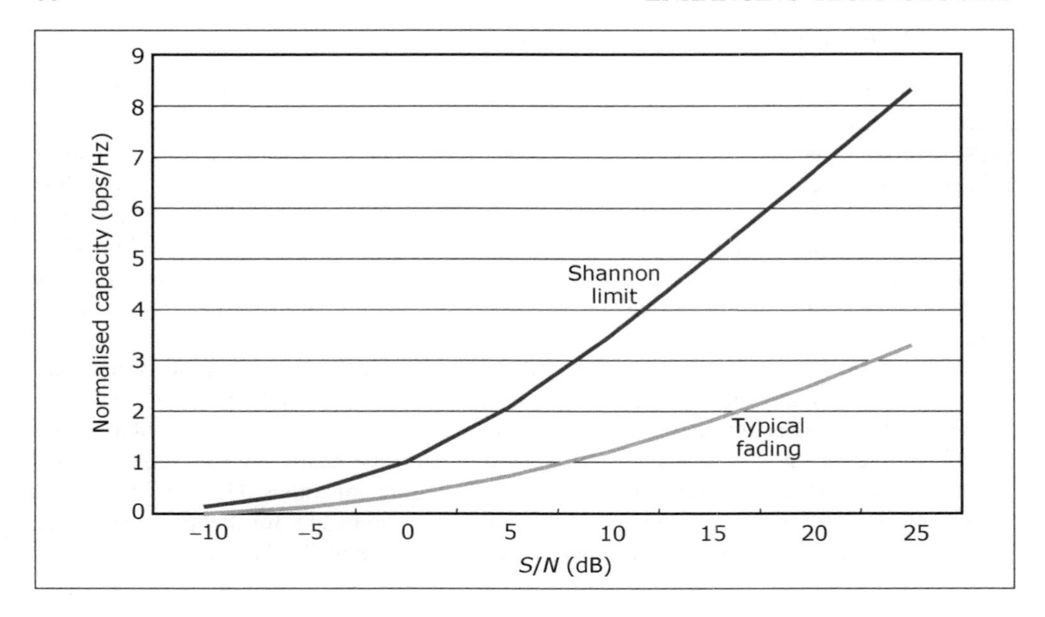

Figure 3.2 Maximum normalised capacity vs SNR, and actual typical throughput

Accordingly, capacity can be increased if SNR can be increased. Figure 3.2 illustrates how normalised capacity (frequency-utilisation efficiency) changes as SNR increases. As noise power is dependant on the bandwidth of the system and the temperature of operation, it generally cannot be reduced. Signal power, however, can be a variable. The total amount of power is limited by several factors, including the device power rating, power amplifier, government regulations, and so on. In any case, the Shannon theorem, as expressed above, applies to noise-limited networks. An approximation for interference-limited networks, which concern most mobile communications systems, is as follows.

3.2.1 Capacity in the presence of interference

Modelling total multi-user interference behaviour as noise is possible if certain conditions such as a minimal number of co-channel interference sources are met. Under such conditions, the Shannon theorem can be approximated to include interference, I, as follows:

$$\frac{C}{B} \approx \log_2\left(1 + \frac{S}{N + I}\right)$$

Thus, any technique that can reduce the interference term, I, will increase system capacity. In the same way, any technique that can increase the signal power term S without increasing I also increases capacity. The enhancing technologies discussed in this chapter are all designed to deliver either of the above – reduced I or increased S.

Figure 3.2 represents the upper limit of a capacity in a noise-affected system. Systems that operate under fading conditions exhibit significantly lower capacity. The capacity of a typical fading system is also shown in the figure.

Table 3.1 ACM levels and required SINR levels for BER = 10^{-3} in a typical fading environment

ACM level	Modulation scheme	Coding rate	Required SINR level (dB)	Throughput bps/s/Hz
1	QPSK	1/3	6.5	0.66
2	QPSK	1/2	7.4	1.0
3	QPSK	3/4	10.8	1.5
4	16-QAM	1/3	8.2	1.33
5	16-QAM	1/2	10.0	2.0
6	16-QAM	3/4	13.6	3.0
7	64-QAM	3/4	16.7	4.5

In order to take advantage of the higher capacity afforded by higher signal to interference and noise ratio (SINR) figures, a technique known as *adaptive coding and modulation* (ACM) is utilised. In this technique, the channel conditions are regularly monitored by a transmitter in order to find the SINR at an intended receiver. On the basis of this SINR value, the most suitable combination of channel coding and signal modulation are chosen to facilitate high throughput. Table 3.1 shows typical SINR levels necessary for several different combinations of modulation and coding rates in fading channel environments. The peak throughput is clearly obtained when the highest ACM level is used and when SINR is very high.

To illustrate and evaluate each of these enhancing technologies, we first explain the sources of signal and interference as below. We will then explain how, and to what degree, the enhancing technologies can improve the performance of a particular system.

3.3 Signal and Interference

In licensed bands, the sources of interference are other co-channel users. Co-channel users may be sufficiently far apart because of frequency reuse, in which case, their interference is designed to be minimal. When the frequency reuse factor is one, which is the case with CDMA systems, then co-channel interference is suppressed through spreading gain. Co-channel use may remain low due to physical distance even if the frequency reuse factor is one for some OFDM systems.

In this section, we examine the signal-to-interference ratios for both the uplink and downlink of wireless systems.

3.3.1 Downlink

The signal to interference-noise ratio in the downlink can be described as follows. The signal term is the power level intended for a particular user. It is a fraction of the transmitted signal from a base station. Denoted as S_{D}, desired signal, it is the received signal power level

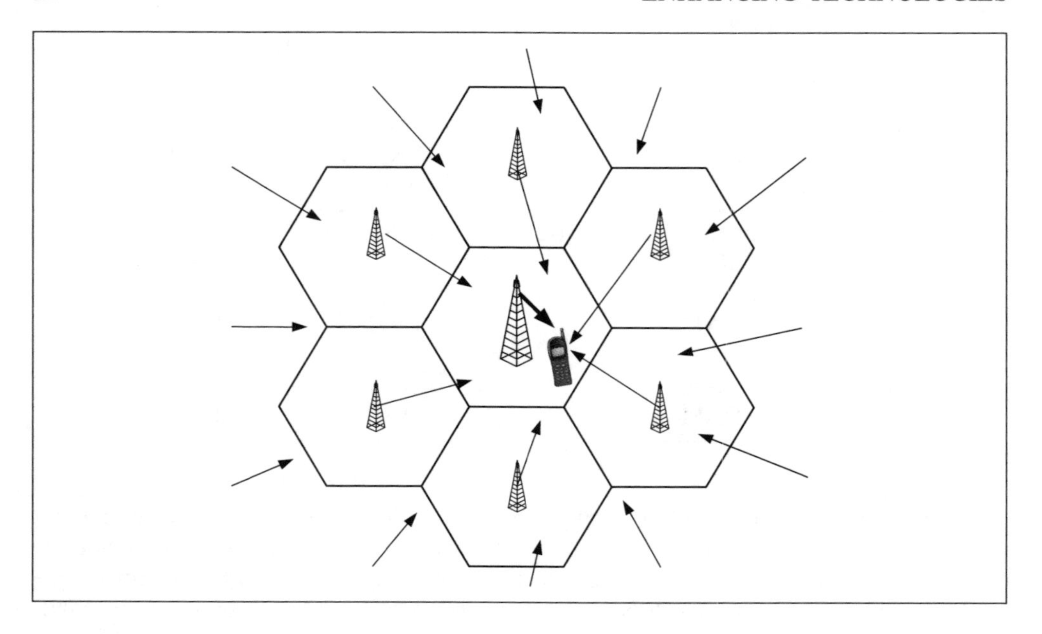

Figure 3.3 Interference in the downlink

at the mobile station, having gone through the transmission channel and its corresponding path loss. Interfering sources are illustrated in Figure 3.3, and can be listed as:

1. Control signals from the serving base station, I_C
2. Signals to other users within the same cell, I_O
3. Multi-path interference, I_M
4. Signals from other BSs, I_B.

All of these terms are interference to the power level as received at the mobile station – the transmitted signals multiplied by their corresponding path loss to the desired mobile station receiver. Finally, there is noise power, N. SINR can be mathematically expressed as:

$$\mathrm{SINR_{DL}} = \frac{S_D}{I_C + I_O + I_M + I_B + N}$$

3.3.2 Uplink

Illustrated in Figure 3.4, and similar to Figure 3.3, the uplink signal to interference-noise ratio can be defined. Here again, the desired signal, s_d, is received from a mobile station, while interfering signals are received from the following sources:

1. Control signals, i_c
2. Signals from MSs within the same cell, i_o

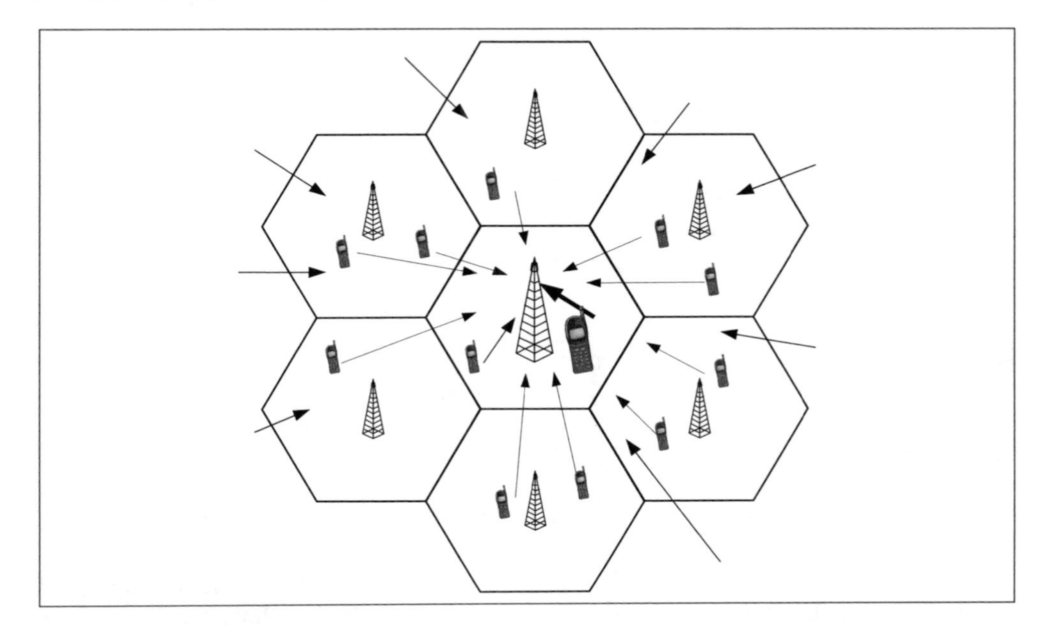

Figure 3.4 Interference in the uplink

3. Signals from MSs within the neighbouring cells, i_b
4. Multi-path interference, i_m.

Again, all these terms denote the power level received at the base station, and are equal to the transmitted power multiplied by the corresponding path loss. With n as noise power, SINR in the uplink is as follows:

$$\mathrm{SINR_{UL}} = \frac{s_d}{i_c + i_o + i_b + i_m + n}$$

3.4 Advanced Antennas

If the received signal power level can be increased without increasing the interference power level, then SINR is higher. This results in improved system capacity as expressed by the Shannon theorem. As discussed before, increasing the transmission power, and therefore the received power, is only possible up to a certain degree. Received power can also be increased through utilisation of more powerful antennas at either the transmitter or the receiver sides. However, merely employing more powerful antennas also increases the interference power level, and does not result in any capacity increase. Antennas should be used in an appropriate way that does not increase interference level.

3.4.1 Directional antennas

Directional antennas confine the radiated power from a transmitter to within a certain arc – sector. Inversely, they also receive power from only a certain sector of the coverage

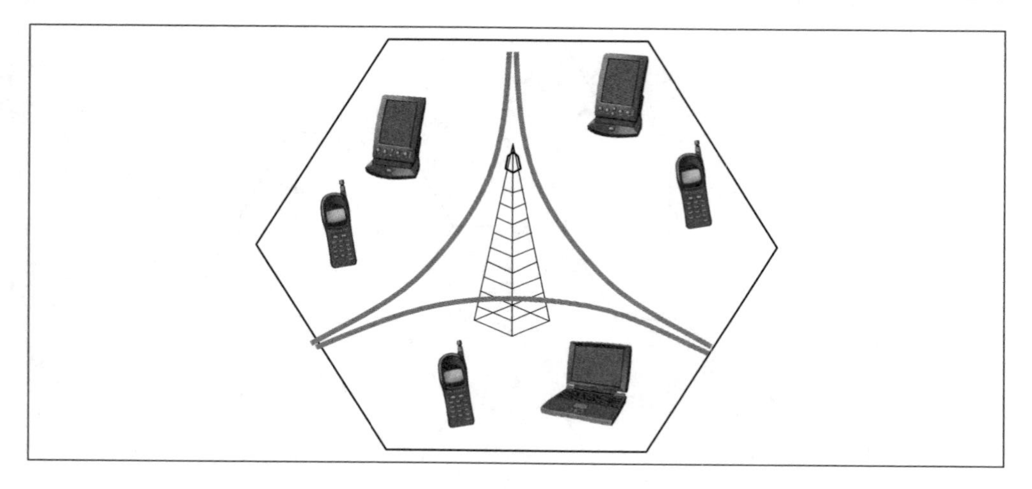

Figure 3.5 A three-sector antenna system

area. Sector antennas used in most 2G and 3G systems are typical examples of directional antennas. Figure 3.5 shows a three-sector antenna system. In downlink, the base station transmits to only a third of its surrounding area, and in uplink, it receives signals only from one- third of the area at each of its antennas. In downlink, we can observe from the equation in Section 3.3 that the terms I_C, I_O, and I_B are all reduced by two-thirds. If I_M and N are ignored, SINR increases by a factor of 3. This results in a significant increase of the throughput within the sector. The usage of sector antennas, coupled with independent transceivers has the effect of multiplying the total system capacity by the sectorisation factor.

In uplink, the usage of sector antennas again reduces interference. Referring to the uplink equation, we can see that i_o is reduced by two-thirds, thus increasing $SINR_{UL}$ and capacity in a way similar to downlink.

The examples above are for a three-sector antenna, with each antenna element having a 120° coverage span. Six-, or even twelve- sector antennas are realisable, and some are indeed utilised. Inevitably, higher orders become impractical because of both cost and installation constraints. In any case, although their use incurs increased cost, sector antennas do increase total system capacity.

Directional antennas are practical only for base stations. Mobile stations, by definition are mobile, and therefore often change directions. The cost of directional antennas is relatively high, which makes them somewhat impractical also for fixed wireless access solutions.

3.4.2 Adaptive array antennas

One problem with sector antennas is that the radiating/receiving patterns are fixed. There is another class of antennas that uses a number of radiating elements, whose outputs are multiplexed to emphasise signals arriving from a desired direction. The multiplying parameters may be adaptively changed to track a moving transmitter. These are known as *adaptive array antennas* (AAAs), and their receiver structure is illustrated in Figure 3.6. The elements may be arranged in an array, either in a circular or in a stack format, as illustrated in Figure 3.7.

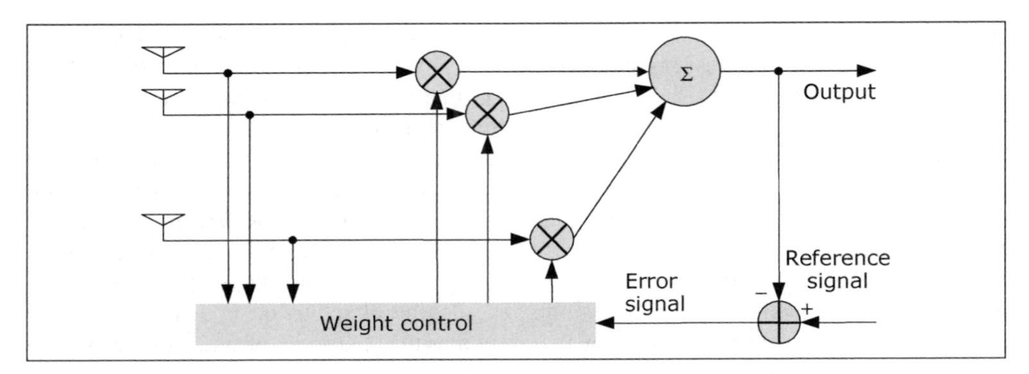

Figure 3.6 Adaptive array antennas

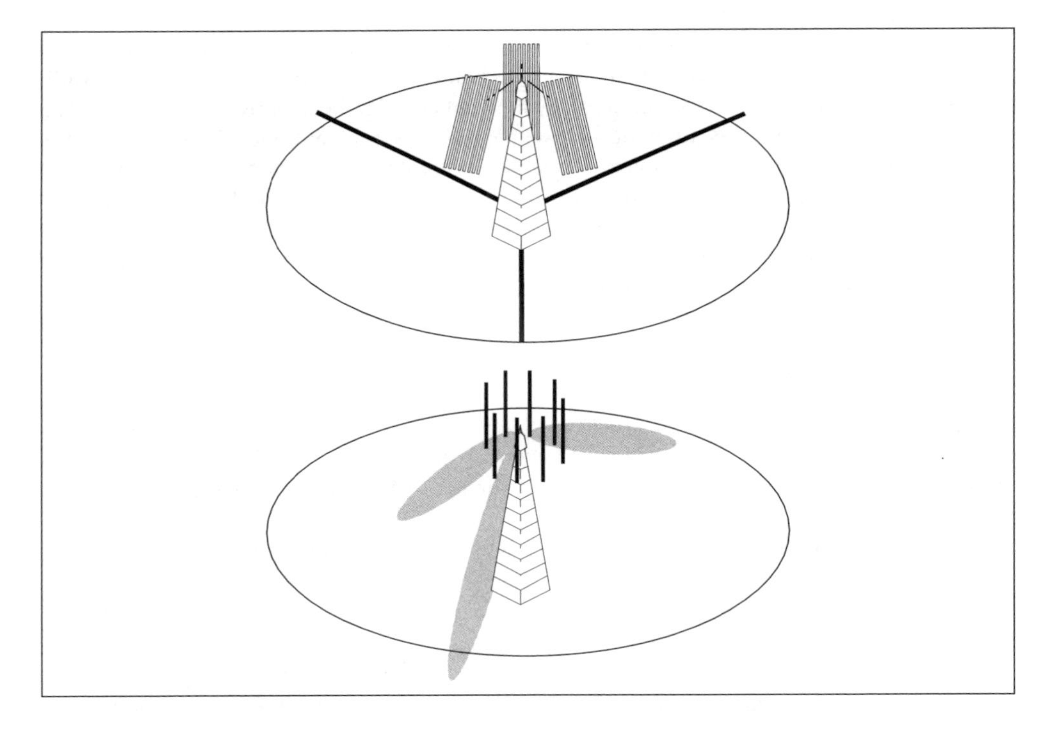

Figure 3.7 Sector and circular adaptive array antenna implementations

Each element of an AAA is connected to a unit which multiplies the receiving or transmitting signal by a complex parameter. By properly setting these parameters, the antenna may be steered towards a desired direction. At the base station, these parameters are calculated and set based upon training sequences sent from a mobile station. The parameters must be regularly updated as a mobile travels within the cell. And, because the parameters are a function of the carrier frequency, the parameters used for uplink need to be calibrated for the downlink transmission. This tends to be nontrivial as other factors such as temperature, and the length of connecting cables, also need to be considered. TDD systems have an advantage over FDD systems in that they use the same carrier frequency for both uplink and downlink and do not require such calibration.

The sharpness of the directed beam is a function of the number of array elements; the more the elements, the sharper is the beam. As with sector antennas, the sharper the beam width is, the more it can reduce interference from co-channel users. Referring again to the formulas of Section 3.3, we can observe that the terms I_C, I_O, and I_B are reduced in the downlink and that i_o and i_b are reduced in the uplink. These result in the reduction of SINR in both the uplink and the downlink. The degree of reduction is a function of the number of antenna elements. A rule of thumb is that the sharpness of an AAA beam, and therefore its ability to reject interference, is approximated by 360° divided by the number of the elements.

As with directional antennas, AAAs coupled with independent transceivers can increase system capacity significantly.

We have already defined in Section 2.5 that antenna elements need to be separated by half a wavelength ($\lambda/2$) for antenna transmission and reception diversity to work well. Similarly, for an AAA to work well, the array elements need to be apart by $\lambda/2$ or more. The exact necessary separation is a function of the channel environment but $\lambda/2$ is a reasonably adequate figure. As the total possible antenna size is limited, the maximum total number of array elements can be calculated from the size as well as the required element separation.

As shown in Table 2.2, operating at higher frequency bands can lead to smaller antenna sizes. This means that array antennas can be deployed in both the base station and end-user equipment sides, leading to further performance improvements.

3.5 Coverage Extension

The coverage of a system, as discussed in Section 2.9, is bound by natural signal decay caused by transmission through the air. A typical model, the Extended Hata model was used to show that path loss can be calculated for relative distances between a transmitter and a receiver.

Coverage is the maximum distance over which a radio signal may be received and detected with sufficient quality. Here, it is the necessary SNR that is of interest. This is because maximum coverage is needed for sparsely populated areas where little co-channel interference would exist from neighbouring cells.

Maximum coverage of a base station can be expanded simply by increasing transmission power (see box Cell radius extension). This will increase the S term in the Shannon theorem capacity equation. This is, however, not a practical solution, because power amplifiers are among the most expensive components in a base station. Furthermore, the same power amplification must occur in the mobile station for the signal to be heard at the base station. Power consumption and device size considerations will rule this option out.

A more practical way to increase the S term is through the CDMA associated process of spreading gain. While this reduces the transmission rate – the lower the transmission rate compared with transmission bandwidth, the higher is the spreading gain – it also increases the effective received signal power S. Table 3.2 shows the possible signal power gain from WCDMA's variable spreading factors.

Cell radius extension

The extra gain from spreading, or from adaptive antennas, can be used to extend the coverage radius. Assuming Hata model of Chapter 2, the radius extension in dB is tabulated as:

Extra gain	Extra coverage
5 dB	2 dB
10 dB	4 dB
15 dB	6 dB

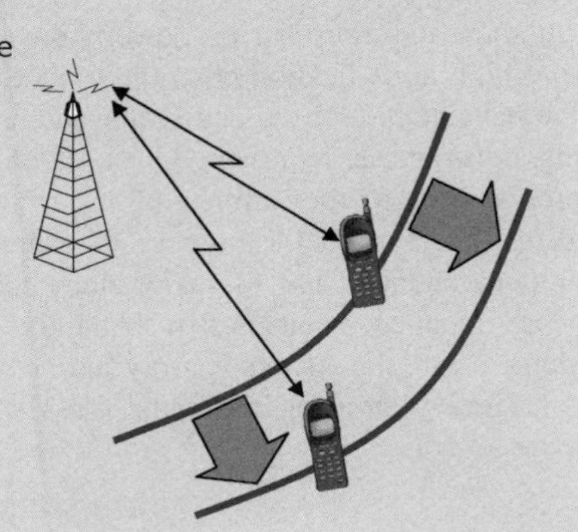

Table 3.2 Spreading gain in WCDMA (chip rate = 3.84 mcps)

Transmission rate (kbps)	Spreading gain (dB)
2000	2.8
1000	5.8
384	10.0
64	17.8
32	20.8

In TDD systems, the cell radius is limited by the guard-time between data slots. The coverage may not be extended beyond this hard limit, even if the transmission power increased (see box "TDD coverage").

TDD Coverage

In a TDD system, time slots are used for uplink and downlink communications. The guard-time between the uplink random access slot and the following time slot determines the maximum possible coverage.

In TD-CDMA the guard-time is 50 μs. This corresponds to one round-trip signaling over the air between the base station and the farthest possible mobile. Therefore, the farthest possible mobile may be located at:

25×10^{-6} (s) $\times 3 \times 10^{-8}$ (m/s) = 7500 m

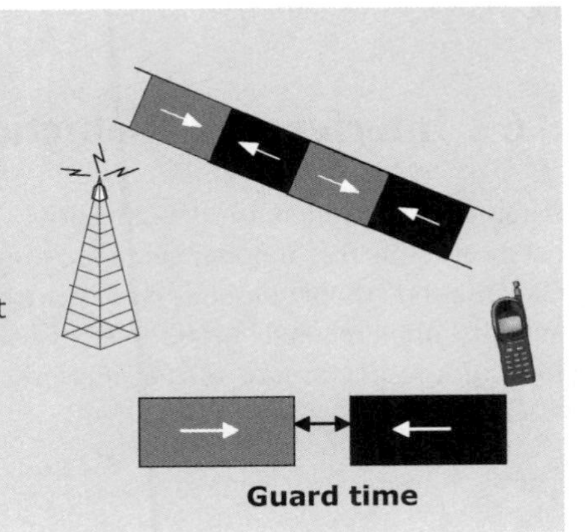

3.5.1 Coverage extension using adaptive array antennas

Although the transmission power from a device cannot simply be increased without incurring high amplification cost, the radiated power to a particular direction can be increased. We have already discussed sector antennas above. AAAs are particularly useful in extending coverage at relatively low cost. As discussed above, the directivity of an AAA is proportional to the number of its array elements. Likewise, the total radiated power is also a function of the number of elements. As one power amplifier is used to feed one antenna element, use of many elements can result in a large increase in received power at an intended receiver. Box 'Adaptive antennas and extension' illustrates how an eight element circular adaptive array antenna can deliver 64 times as much power (18 dB) to a desired user using the same class of power amplifiers as compared to a single element antenna.

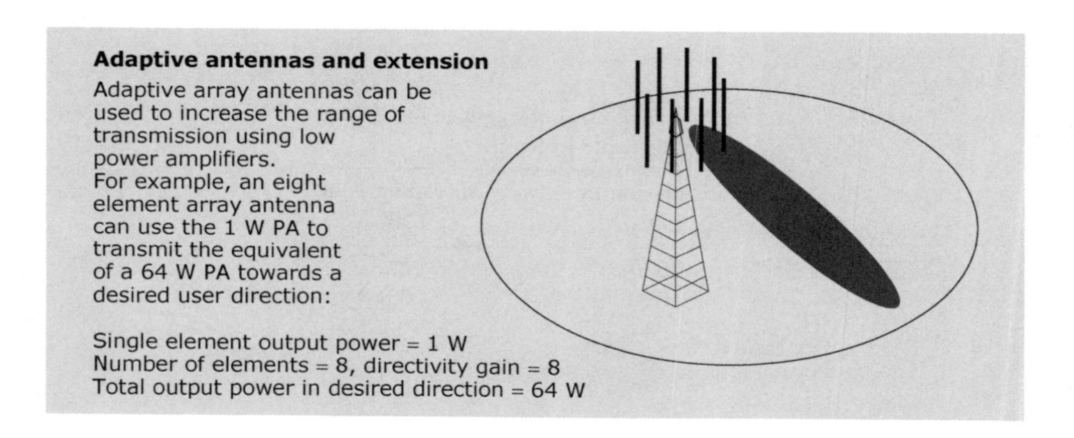

While CDMA systems use both the spreading process and AAAs to extend their coverage, TDMA based systems can only rely on AAAs. The consideration of maximum coverage is an important parameter in the business plan of an operator and is a factor in the choice of technology.

3.6 Interference Reduction

Another way to increase the capacity of a system is to reduce one or more of the interference terms in the denominator of the equations in Section 3.3. A number of interference cancellation (IC) techniques were devised for 2G and 3G systems, although very few were actually implemented. In 4G systems, the technology is being designed in a suitable way to minimise occurrence of the interference.

3.6.1 Interference cancellation

Of the few existing interference reduction techniques, IC is based on simple arithmetics. An interferer user's signal may be detected, calculated and then removed, as shown in Figure 3.8. This is generally possible for all FDMA, TDMA, and CDMA systems. However, as the number of co-channel users increase, the signal processing requirements become prohibitively larger than the processing capabilities of the 1G and 2G systems at the time of their deployment. Moreover, as the gains provided by these IC techniques do not warrant the extra processing costs, these systems were not commercially implemented in 1G and 2G systems. In particular, the IC technique was investigated for GSM systems, but was not commercialised.

IC was more seriously studied for 2G and 3G CDMA systems. The reason was because of these systems' frequency reuse factor of one, and the fact that all users transmit in the same frequency band. Initially, IC techniques were deemed promising. However, again the performance enhancement gain was not high enough to warrant commercial implementation.

In the downlink, IC techniques can theoretically reduce or totally remove the I_O, I_M, and I_C terms of page 62 equation. However, this generally comes at the expense of significant extra processing, and signalling from a BS to an MS, which make their implementation impractical, if not impossible. IC techniques were more seriously studied for the uplink

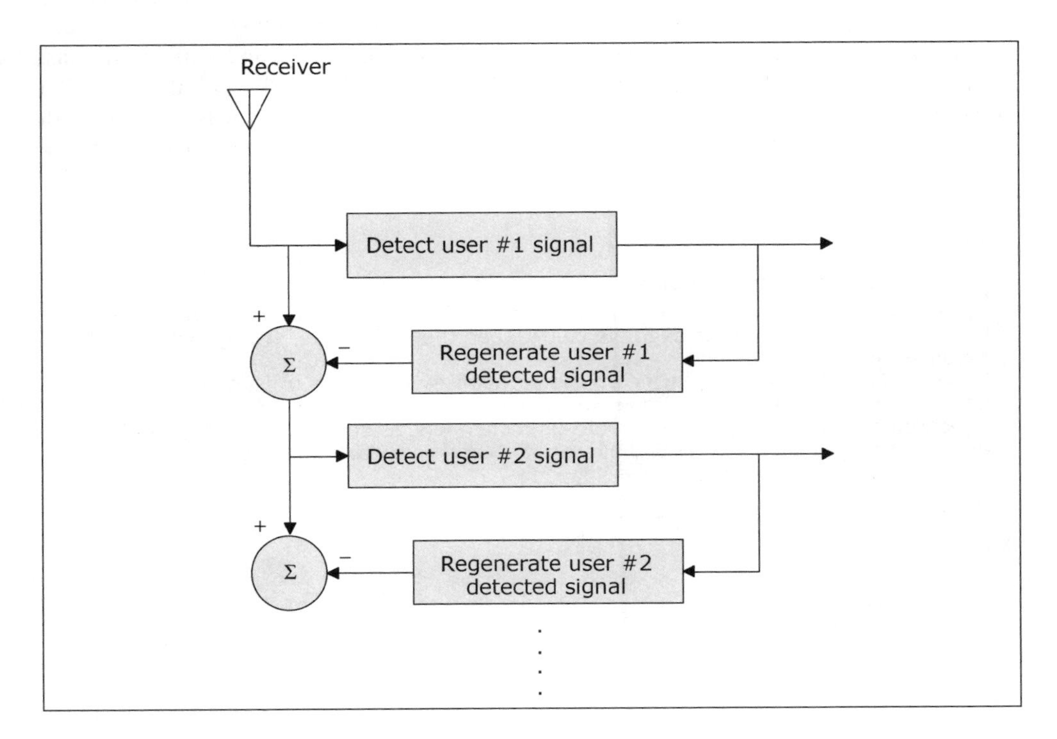

Figure 3.8 Serial interference cancellation

receptions at a BS, where more signal processing power can be supplied, as opposed to a relatively simple MS. Here, IC techniques can theoretically remove the r_o and possibly the r_m terms of the page 63 equation. The maximum benefits of these systems cannot, however, be realised in practical system operating conditions. Even suboptimal performance comes at the expense of too large processing requirements. All of these meant that IC techniques never reached the commercialisation stage. Throughput improvement using IC can be found from Figure 3.2: any increase in SINR leads to an increase in system capacity using the typical fading curve. For example, for 5 dB improvement in SINR of a system operating at SINR = 5 dB yields a 50% increase in throughput. The increase in SINR is a function of channel conditions as well as processing capability. Typically, a decrease of 2–3 dB can be expected.

3.6.2 Joint detection

Joint detection (JD) or multi-user detection (MUD) is another class of interference reduction technique. JD is used in CDMA systems, and is based on processing in the code domain. In CDMA systems, it is possible to reduce interference by considering the correlation properties of the spreading codes. In CDMA systems, each user communicates using a distinct spreading code by which the user can be distinguished at the receiver side. The signals of co-channel users are mutually interfering as stated above. However, interference may drop to zero if synchronous transmission with orthogonal spreading codes are used.

Synchronised orthogonal transmission does not work perfectly under all conditions. Firstly, it is only useful for users in the same cell and in the downlink (with the exception of TD-SCDMA). Moreover, it is of maximal benefit in flat fading channels; in multi-path channels, orthogonality is reduced and therefore multi-user interference is increased (see Box: Synchronous transmission and orthogonal spreading codes).

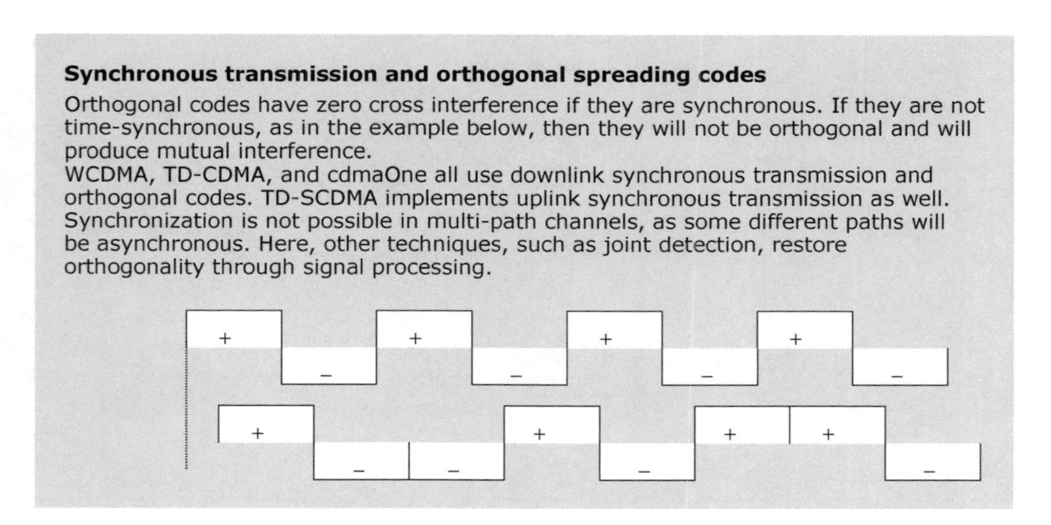

Synchronous transmission and orthogonal spreading codes

Orthogonal codes have zero cross interference if they are synchronous. If they are not time-synchronous, as in the example below, then they will not be orthogonal and will produce mutual interference.
WCDMA, TD-CDMA, and cdmaOne all use downlink synchronous transmission and orthogonal codes. TD-SCDMA implements uplink synchronous transmission as well. Synchronization is not possible in multi-path channels, as some different paths will be asynchronous. Here, other techniques, such as joint detection, restore orthogonality through signal processing.

JD is the process by which the signal of several users is jointly detected considering their spreading codes' cross-correlation properties. In this process, the co-channel user interference caused by sub- or nonorthogonal spreading codes is removed. This is the

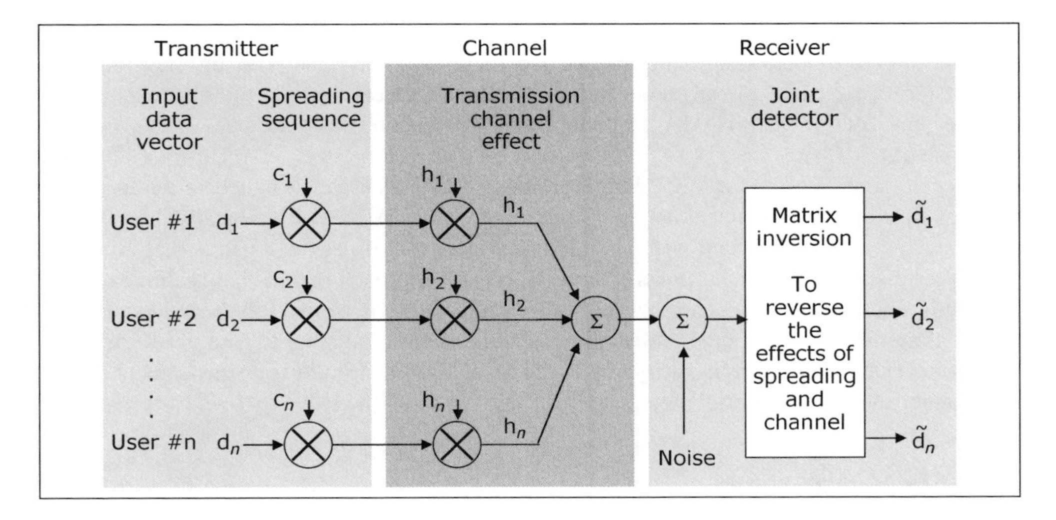

Figure 3.9 Joint Detection

process of code/multi-user domain equalisation referred to in Chapter 2. Through a perfect JD process, orthogonality is restored. In practice, JD is a matrix inversion and multiplication process, as illustrated in Figure 3.9, where the matrix inversion is effectively equalisation in the code domain. In CDMA systems, the JD process restores orthogonality in both downlink and uplink, and in the presence of multi-path interference.

It should be noted that orthogonality is also restored through the IC process defined above through actual removal of interference.

The size of the above matrix inversion calculation is a function of the size of spreading codes and the number of interfering users. Both 2G and 3G voice CDMA systems are designed to accommodate a large number of users at the same time. This increases the complexity of a JD receiver at the BS, again making it impractical for actual implementation. The situation is even more difficult in the downlink. However, the 3G TD-CDMA standard was designed specifically to use JD. In this standard, the number of users at each instant is restrained, and the spreading code length is also kept small at 16. This means that JD can be implemented in both downlink and uplink. Until recently, JD systems had been designed to reduce the interference generated by the users within the same cell, and remove the interference caused by multi-path. With faster processing, it is now possible to equalise even for inter-cell interference. This will further increase frequency-utilisation efficiency.

The JD process with inter-cell interference reduction removes the I_O, I_M, and I_C terms and part of the I_B term in Section 3.3 equation for the downlink. In the uplink, the i_o, i_m, and i_c terms, and part of the i_b term in the uplink equation can be reduced. Capacity improvement may be calculated in a similar way as in Section 3.6.1. With inter-cell interference reduction, an SINR improvement of 5 dB is possible.

3.6.3 Interference avoidance

It is possible to design systems in such a way that some of the interference terms in the equations in Section 3.3 disappear. We have already discussed OFDM systems in Chapter 2.

The sub-carrier bandwidth parameters for these systems can be chosen in such a way that the effects of multi-path interference are removed. This results in the near elimination of terms I_M and i_m from these equations. The co-channel users within the same cell can be removed using a TDMA approach: this will ensure that terms I_O and i_o become nonexistent.

Indeed, such systems as WCDMA HSDPA and WiMAX are designed to avoid one or both of these two particular sources of interference. WiMAX carrier bandwidths will be quite narrow. This, combined with a sufficiently large guard period, ensures that no multi-path interference is generated. Furthermore, a TDMA approach means that within each cell area only one user is active, and therefore, no multi-user interference is generated. As co-channel interference from other cells still exists, there does remain a degree of interference. The same TDMA approach is partly employed in WCDMA HSDPA transmissions to reduce co-channel multi-user interference.

3.7 Hybrid ARQ

Although not an enhancing technology in the same sense as discussed above, HARQ is a practical technique that can enhance the throughput of a system. It works together with the ACM technique discussed in Section 3.2. Automatic repeat request (ARQ) is a particular technique for packet switched services, for which very low packet error rates are essential. In ARQ technology, the receiver checks for errors in each received packet and, if an error is detected, a request for re-transmission of the packet is sent to the transmitter. This process is illustrated in Figure 3.10. ARQ is applied to both fixed and wireless communications.

In fixed communications, the occurrence of an error is a very rare event, and when it does happen, the erroneously received packet is usually discarded, thus triggering ARQ processing through higher layers. Most ARQ events in fixed-line communications result from dropped packets, that is, packets that do not arrive at all. The occurrence of a dropped packet can be known because its sequence number will be missing at the receiver.

In wireless communications, however, a packet error event is not rare and occurs with a typical probability of 10^{-2}. An erroneously received packet is not discarded because it still carries some information. In this case, an ARQ is still made, and the re-transmitted packet is combined with the initially received packet to enhance probability of correct reception. This technique is known as *hybrid ARQ* (HARQ). The following two HARQ methods are specified in 3GPP standards. They are applicable to other broadband wireless technologies as well.

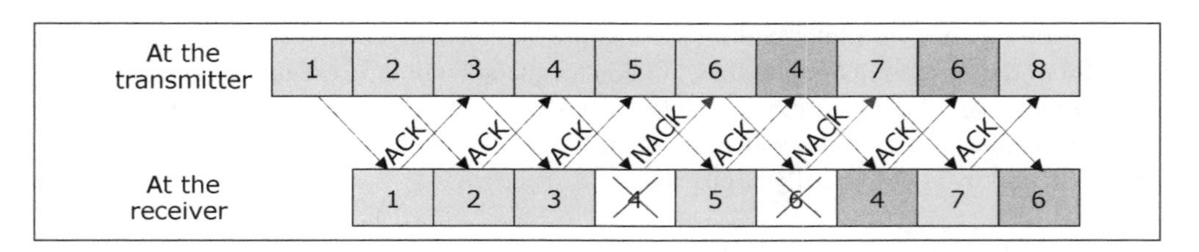

Figure 3.10 Automatic repeat request (ARQ). ACK and NACK are messages sent back to the transmitter, informing correct reception (ACK) or incorrect reception (NACK)

3.7.1 Chase combining

In the Chase combining technique, an erroneously received packet is combined with its re-transmitted replica as illustrated in Figure 3.11. Chase combining technique facilitates data transmission even in very low SINR environments. The signal combining process of the HARQ method increases the SINR with each subsequent reception of each packet.

3.7.2 Incremental redundancy

The incremental redundancy (IR) technique takes advantage of variable FEC coding rates. In the IR technique, a low coding rate is initially chosen for transmission. However, some of the redundancy code bits are punctured (i.e. not transmitted). As puncturing reduces the capacity of an FEC code to correct errors, the probability of error is increased. If the packet is received without errors, then a higher frequency-utilisation efficiency is obtained. If the packet is received with error, then the punctured bits are sent, restoring the code's original correction capability. The operation of an IR system is illustrated in Figure 3.12. Modes in the figure refer to those of Table 3.1.

Figure 3.13 illustrates the throughput of a HARQ system. It is interesting to compare this figure with the Shannon limit capacity of Figure 3.2, and to note how similar the

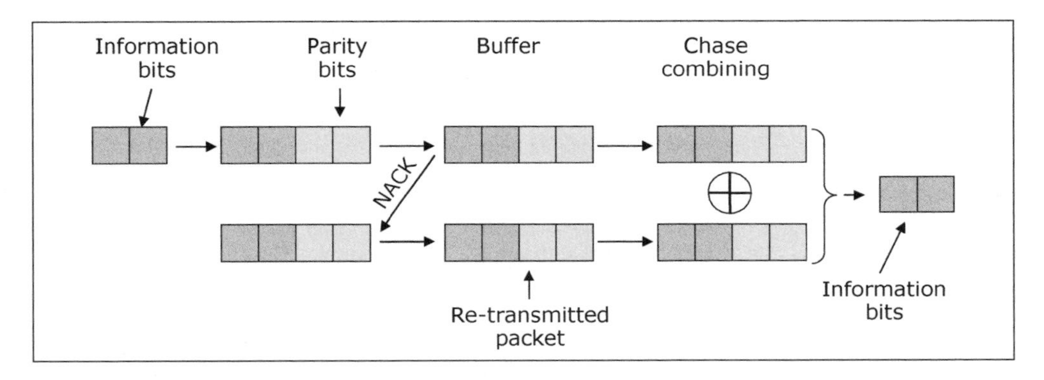

Figure 3.11 Chase combining

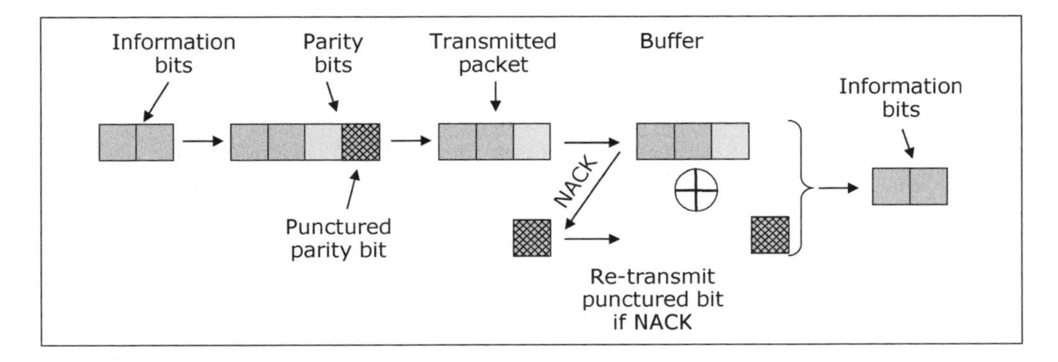

Figure 3.12 Incremental redundancy

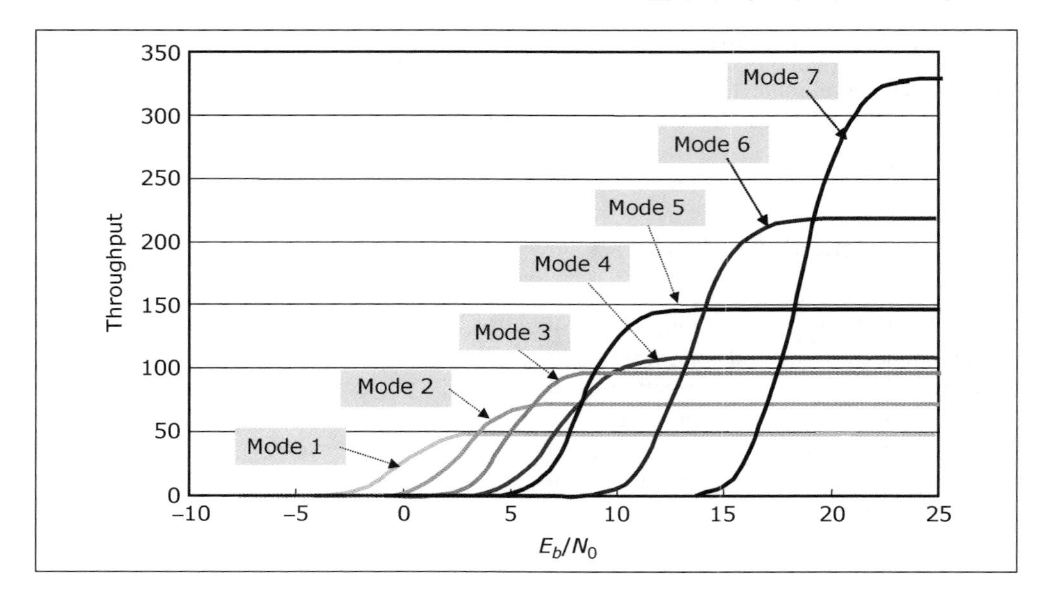

Figure 3.13 Hybrid ARQ throughput

two figures are. The lower curve in Figure 3.2 is in fact the upper envelope of HARQ performance from Figure 3.13.

3.8 MIMO Antennas

A MIMO antenna system is a recent technology development. Based on the work by Alamouti in 1998, this technique has been the focus of much research and has already been adopted in several standards, including 3GPP and IEEE 802.11.

The technical details of this technology are beyond the scope of this book. However, the effect this class of antenna has is to significantly increase the capacity and throughput of a wireless communication system. In effect, the MIMO antenna technology creates parallel channels and results in an increase in the total throughput of the system.

Figure 3.14 illustrates a MIMO system. On the transmitter side, there are M transmitting antenna elements, and on the receiver side, there are N receiving antenna elements. Assuming $N < M$, theoretically this system has an N times throughput as compared to a transmitter/receiver system with single antenna elements. MIMO can further introduce transmission diversity to a system where other forms of diversity cannot be available. Table 3.3 lists the advantages of a MIMO system in introducing diversity and increasing capacity for a number of different transmitter/receiver antenna elements.

The ability to increase system throughput with a simple addition of antenna elements to the transmitter and receiver is of course of great interest to a wireless operator. There are, however, some conditions. First, the antenna elements need to be sufficiently separated for the full MIMO capability to be realised. Again, typically a half-wavelength ($\lambda/2$) antenna-element separation is considered necessary. While this may be possible at a base station, it may be difficult to realise at a mobile station. Operating at higher carrier frequencies reduces

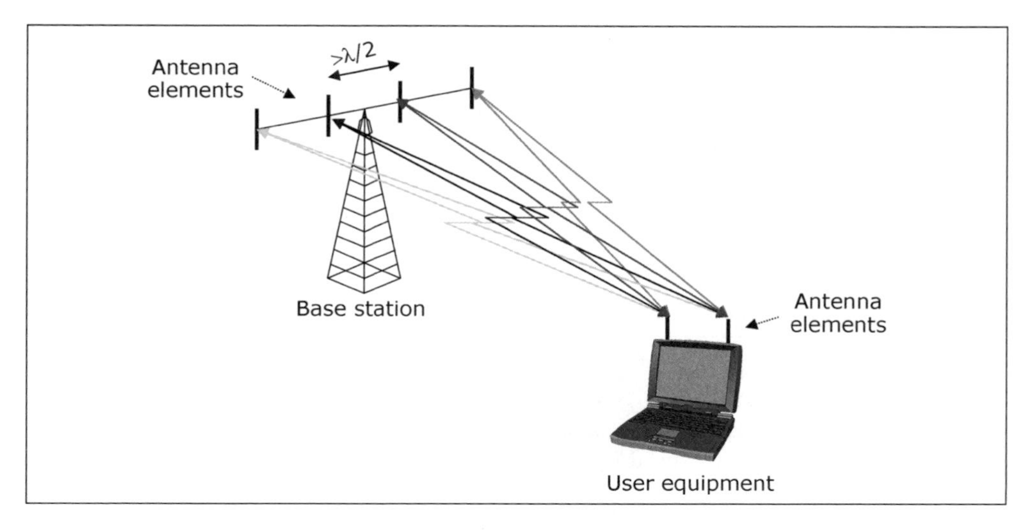

Figure 3.14 A MIMO system

Table 3.3 Diversity and capacity improvement using a MIMO antenna system

Number of elements at base station	Number of elements at end-user device	Effect
2	1	2 × diversity
2	2	2 × diversity, 2 × capacity
4	2	4 × diversity, 2 × capacity

$\lambda/2$ and makes MIMO realisation more practical. MIMO systems also require extra signal processing. This should not be a concern with faster processors that will be available in future systems.

3.9 Voice Coding

A major complaint with mobile telephony in comparison with fixed lines is the poorer level of received voice quality as perceived by the end-user. The quality of the received voice, in terms of how noisy, jittery, or otherwise distorted, has become known as *perceptual quality of service* (PQoS). PQoS is measured in units of mean opinion score (MOS). MOS is a parameter between one and five, with five representing Excellent quality and one representing Bad quality. MOS values of 3.0–3.5 or above are usually considered as a minimal requirement for a voice operator. MOS has been used by voice (fixed or mobile) operators for many years to measure how good the quality of their services are. Studies by ITU (see Table 3.4) indicate that a low MOS will result in end-users discontinuing their conversations early.

Table 3.4 Early call termination probability caused by poor voice quality (source: ITU)

Quality	Early call termination probability (%)
Excellent	0
Good	0
Average	10
Fair	40
Bad	95

Table 3.5 Some narrow-band and wide-band AMR classes and their MOS values. Reproduced by permission of John Wiley & Sons, Ltd

NB-AMR		WB-AMR	
Rate (kbps)	MOS	Rate (kbps)	MOS
4.75	2.8	6.60	3.2
6.70	3.1	8.85	3.5
7.40	3.2	12.65	4.0
10.20	3.3	15.85	4.1
12.20	3.4	18.25	4.2

In digital communications, voice is converted from analogue to digital. The resulting bit stream is then compressed in order to remove largely redundant information. This process can be highly efficient; while a pure analogue to digital conversion of voice yields a 64-kbps data stream, voice compression technique can reduce this to less than a few kbps. This is an important reason why digital 2G systems and beyond have a better frequency-utilisation efficiency in comparison with the analogue 1G systems.

Voice compression, however, results in distortion of vocal information and, therefore, a reduction in MOS. One voice encoder technology, adaptive multi-rate (AMR) has been standardised for 3G services. Two classes of AMR encoders exist. One is narrow-band adaptive multi-rate (NB-AMR), which has eight different compression ratios, resulting in the compressed bit rates of 4.75 to 12.2 kbps. The other is wide-band adaptive multi-rate (WB-AMR), with nine compression ratios of 6.6 to 23.05 kbps. WB-AMR was introduced to enhance the quality of delivered voice and improve MOS. It is expected that WB-AMR can deliver voice quality on par with fixed telephony. Some NB- and WB-AMR vocoders and their corresponding MOS values are listed in Table 3.5.

PQoS is also relevant to other perceptually significant services such as audio streaming and, in particular, video streaming, and video telephony. MOS standards are being developed by ITU (see Box: Video PQoS).

Video PQoS

Perceptual QoS is defined as the quality of a received signal intended for viewing or hearing, as *perceived* by the end-user. While voice PQoS have existed for a long time, audio and video PQoS are relatively new fields.
A company active in developing PQoS metrics and tools is Genista Corporation. Genista tools such as *Optimacy* measure and evaluate the quality of received video in terms of perceptual SNR, blurriness, jitter, and blockiness. These metrics can then be used to dynamically control video quality. See also Further Reading list at the end of this chapter.

Summary

In this chapter, we have discussed a number of technologies that can be used to enhance the performance of a particular wireless communications technology. We discussed antenna technologies that can be used to increase a system's capacity and maximum coverage, almost independently of the system's technology. The MIMO antennas can generate parallel transmission channels, thereby further increasing system throughput. Another system-independent technique is HARQ, which takes advantage of a transmission environment to deliver higher throughput.

We also discussed system-dependant techniques. These techniques work on specific characteristics of a technology to increase signal power, or decrease interference power and thereby increase the system's capacity. Here, we discussed interference reduction techniques as used in CDMA and TDMA systems and interference avoidance techniques as used in OFDM-based systems.

A combination of these techniques are used in present systems, and are expected to be increasingly used in future systems. Capacity is one of the most important parameters of system design, and technologies that increase the maximum system capacity are among the most valuable to an operator.

Further Reading

- On Adaptive array antenna:
  - El-Zooghby, A., *Smart Antenna Engineering*, Artech House Publishers, 2005.
  - Sarkar, T., Wicks, M. C., *et al.*, *Smart Antennas*, Wiley-IEEE Press, 2003.
- On Multi-user detection, joint detection, interference cancellation:
  - Verdu, S., *Multiuser Detection*, Cambridge University Press, 1998.
- On ACM:
  - Holma, H., Toskala, A. (Editors),*WCDMA for UMTS: Radio Access for Third Generation Mobile Communications, 3rd Edition*, John Wiley & Sons.

- On ARQ:
  - Lin, S., Costello, D., Miller, M., Automatic-repeat-request error-control schemes, *IEEE Communications Magazine*, December 1984.
- On Power amplifiers:
  - Cripps, S. C., *RF Power Amplifiers for Wireless Communications*, Artech House Publishers, 1999.
- On MIMO:
  - Jankiraman, M., *Space-Time Codes and MIMO Systems*, Artech House Publishers, 2004.
  - Larsson, E. G., Stoica, P., *Space-Time Block Coding for Wireless Communications*, Cambridge University Press, 2003.
- On OFDM, see Further Reading list in Chapter 2.
- On PQoS:
  - book by Winkler, S., *Digital Video Quality: Vision Models and Metrics*, John Wiley & Sons, 2005.
  - Genista Inc. web site `http://www.genista.com/`
  - On AMR codec rates for 3G system: 3GPP standard: TS26.110, TS26.171.
  - On early call termination probability: R-Model in ITU recommendation G-107, The E-model, a computational model for use in transmission planning.
- On Adaptive multi-rate vocoder:
  - 3GPP Mandatory speech codec speech processing functions, AMR Speech Codec, 3G TS 26.071.
  - 3GPP Speech codec processing functions, AMR Wideband Speech Codec, 3G TS 26.171.

4

Cellular Topologies

Wireless communication systems provide connectivity to mobile user equipment over a wide geographic area. As stated earlier, it is impractical to provide 100% coverage over an entire country: it is even unnecessary, as there are many uninhabited areas. It is even impractical to provide 100% coverage of populated areas because of return-on-investment concerns. Figure 4.1 shows the areas of coverage for KDDI's CDMA system. The present network provides coverage to some 99.7% of the Japanese population.

For highly populated areas, it is important to provide 100% coverage. However, it is very difficult to do so here. Physical objects – both natural, such as hills and valleys, and man-made, such as buildings, covered areas, tunnels, subways, and so on – block the propagation of radio signals, resulting in insufficient received signal power. An operator designs a network layout to provide full coverage with minimal infrastructure cost. Network planning, a process whereby the location for installation of a base station, the directionality of its antennas, and its transmit power levels are decided, is one of the most important aspects of the business.

Networks are planned using several cell design topologies. Wide-coverage base stations, usually with antenna heights in the range of 30 to 50 m, form the backbone of coverage of large cells. These are known as *macro-cell base stations*. These are complemented by micro-cell base stations which cover smaller areas, and fill in the gaps left within macro-cells. There are still smaller coverage areas, known as *pico-cells*, which fill in the left-over gaps. Other elements such as repeaters, complement these to provide near full coverage, as illustrated in Figure 4.2.

Coverage can be considered from a user's point of view also. A model, developed by Wireless World Research Forum (WWRF), defines wireless networking needs based on an individual user. This user is considered to have connectivity needs on a personal-area, immediate-area, and wide-area basis as shown in Figure 4.3. This approach to network topology design helps bring together the device and network design fields.

This chapter lists and discusses possible network topologies for broadband wireless communications systems. We discuss the characteristics of these topologies and how they perform, considering the specific challenges of broadband systems. These include broadband

Broadband Wireless Communications Business: An Introduction to the Costs and Benefits of New Technologies Riaz Esmailzadeh

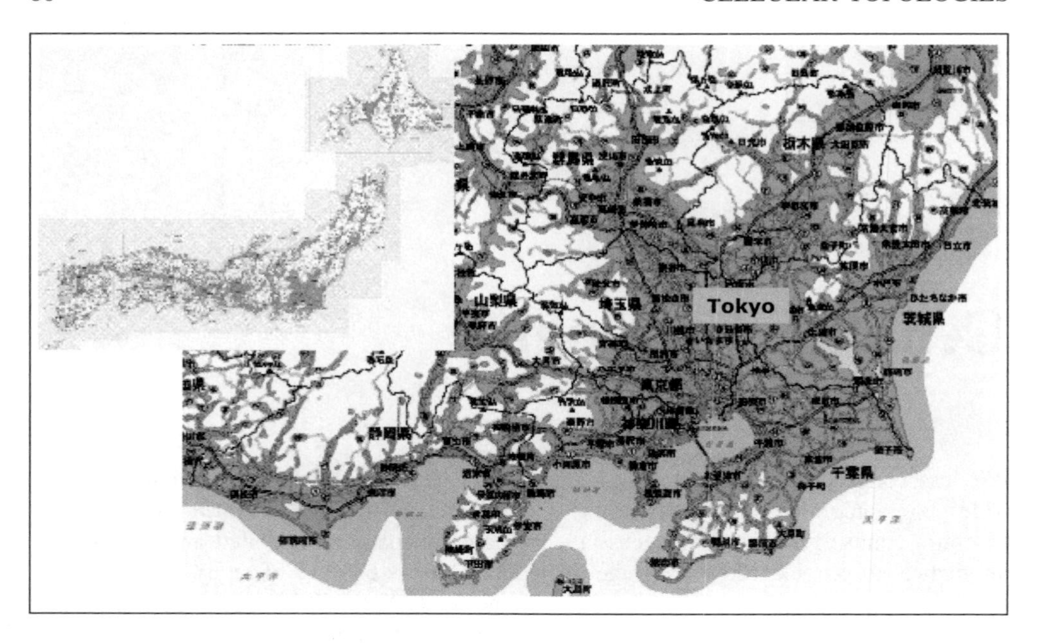

Figure 4.1 KDDI Japan CDMA coverage in Kanto area as of August 2005. Reproduced by permission of KDDI Inc.

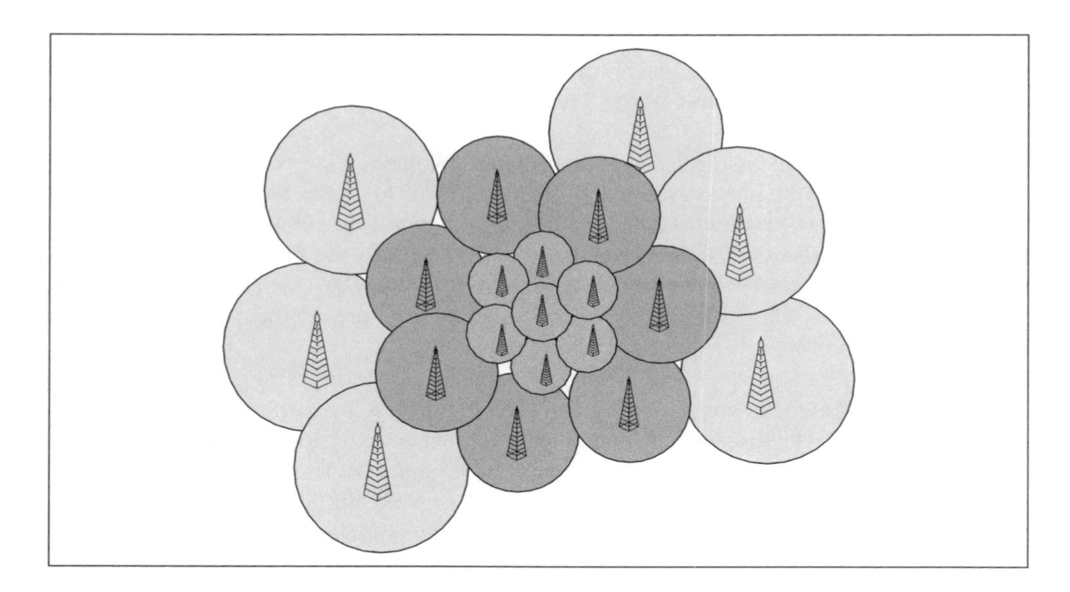

Figure 4.2 Coverage through multi-level cells

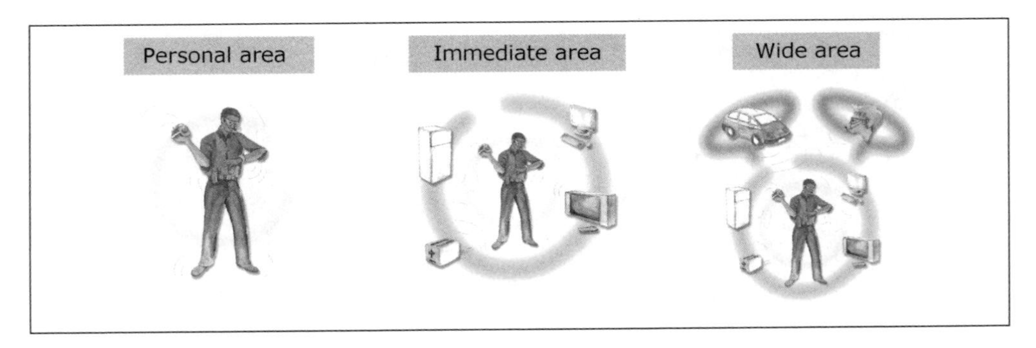

Figure 4.3 Wireless World Research Forum networking model. Reproduced by permission of WWRF

systems' very high transmission rate requirements, as well as challenges posed by operation in the higher carrier frequencies of 2–5 GHz band.

4.1 Cell Structure

The classic style of network design is based upon a single base station, covering a circular area with the base station as its centre. This structure is complemented and enhanced using the technologies described in Chapter 3, including sector and adaptive antennas. As shown in Figure 4.2, several cell structures can be defined on the basis of mainly the complexity of the base station and the size of the coverage area. Here, we define macro-, micro- and pico-cell structures.

4.1.1 Macro-cell

Although there is no standard definition for a macro-cell, this term generally denotes an area with a radius of several hundred metres to a few kilometres, covered by a single base station. Figure 4.4 shows two typical macro-cell base station installations. The base station usually has powerful antennas with a height of 30 to 50 m, installed atop a tower or a tall building. Transmit power can be in the order of up to tens of watts (typically between 5 to 10 w: 37 to 40 dBm).

Macro-cell design topology is used in almost all cellular wireless communication systems. The exceptions are perhaps PHS systems in Japan, and China, although the cell structure here is still sometimes termed as *macro-cell*, despite their coverage range of a few hundred metres.

Macro-cell structure is expected to be the main element also for broadband wireless systems, whether fixed or mobile. Indeed, initial system layouts and system trials use macro-cell configuration. Furthermore, as the number of subscribers to a new system is low at first, it does not make sense economically to provide high-throughput coverage over an entire area using more detailed topologies such as micro- and pico-cells. As we will discuss in Chapter 7 the initial costs as well as operating costs are a function of the number of cell

Figure 4.4 Two macro-cell base stations

sites. As smaller cell topologies will require a larger number of sites, higher initial costs can be avoided using macro-cell structure. Subsequent introduction of micro-cells may be made in targeted areas as necessary.

The macro-cell radius depends on two factors: the amount of expected offered traffic at peak time, and the maximum coverage of the technology. Expected offered traffic is a function of population density. Therefore, in crowded urban areas, the cell radius is usually very small. In suburban and rural areas, population density is relatively smaller and therefore a base station can cover a larger area, which means the cell radius is larger. The height of base station antennas are a function of cell topography and urban building characteristics. Typically, four types of macro-cell designs can be defined:

1. Dense-urban: Downtown in major cities, characterised by high-rise building and high population density. The cell radius can be in the order of 500 m. Typical antenna heights are 35–50 m. Cell sites are usually rooftops.

2. Urban: Town-centres, characterised by medium-height building. Cell radius is in the order of 1 km and base station antenna heights are 25–40 m. Cell sites are usually rooftops.

3. Suburban: Urbanised areas, with single- to multi-storied buildings. Cell radius is in the order of a few kilometres, and antenna height is 10–25 m. Cell sites are usually rooftops or specially built towers.

4. Rural: Sparsely populated areas with little traffic. Here, the cell radius is determined by a system's link budget, that is, how far the signal can travel before received signal-to-noise ratio falls below a required threshold. The cell radius also depends on the technology's coverage limitations.

4.1.2 Micro-cell

The size of a micro-cell can be defined as being in the order of one-tenth to one-twentieth of a macro-cell. Micro-cell design and base station installation come at a later stage in the network plan. A micro-cell base station generally has a less powerful, smaller antenna. Its transmit power is also smaller than a macro-cell's by 10 dB or more. It may be connected directly to a central network, or it may be connected via a macro-cell's base station. Figure 4.5 shows a micro-cell base station.

One reason that micro-cells are added is to fill in the gaps in a macro-cell based network structure. They are installed in areas where coverage cannot easily be provided using macro-cells. These can exist in highly built-up urban areas, where high-rise buildings block the signals and create areas with low signal power levels. They can also exist in isolated areas such as underground train stations or shopping areas. Another reason for introducing micro-cells is for increasing throughput in the so-called *hot spots*. This is accomplished by increasing the SINR level in a particular area (see Chapter 3), and thereby increasing system capacity and throughput.

Figure 4.5 A micro-cell base station

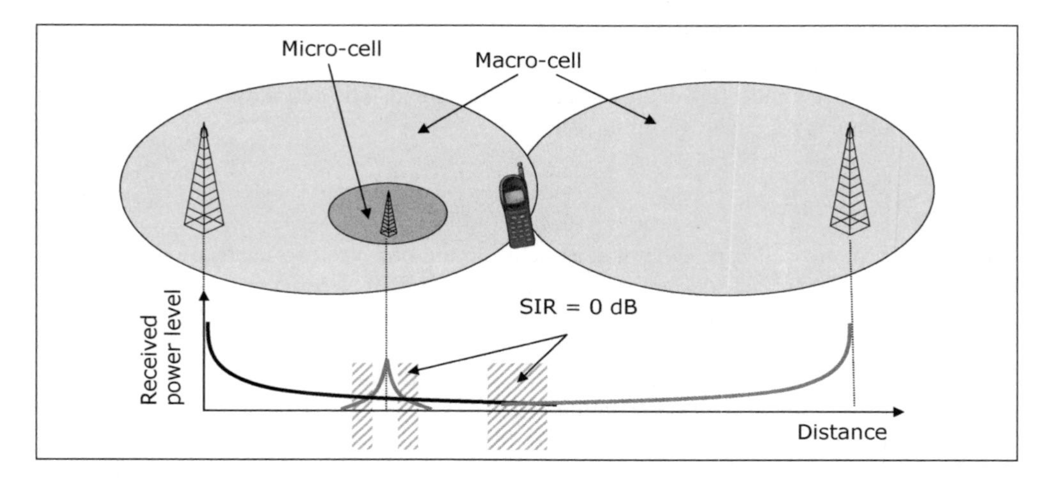

Figure 4.6 Mutual interference in macro- and micro-cell structures

A micro-cell may underlay an existing macro-cell service area, or it may complement the original macro-cell service area by providing service where a macro-cell base station does not reach. In any case, two base station signals will interfere with each other. The network planner's job is to minimise interference caused by this coexistence. Figure 4.6 shows the received power level from two macro-cells and a micro-cell. As expected, the received power is highest in areas near the base station (for both macro- and micro-cell); however, there do exist areas where the received power from macro- and micro-cell base stations are equivalent.

If frequency reuse factor is one, such as in CDMA systems, then one of the signals of Figure 4.6 becomes an interference to the other. This means that signal-to-interference ratio is one (or zero dB), which translates into very poor signal quality (shaded areas in Figure 4.6). For this reason, CDMA systems use spreading gain to increase the level of desired signal through de-spreading. TDMA and FDMA systems (including pure OFDM) do not spread signals and therefore do not have any way of finding extra gain at cell edge. As a result, they cannot use the same frequency carrier in neighbouring cells: that is, they have frequency reuse factor of more than one. It should also be mentioned that in CDMA systems, macro- and micro-cell overlay topologies are difficult to design, as SIR can be very low at the edge of overlapping cell areas as illustrated in the figure. In these systems, micro-cells are added in electromagnetically isolated areas as defined earlier.

In systems where the carrier frequency reuse factor is larger than one, such as GSM, the overlapping macro-cell and micro-cell base station carrier frequencies can be, and usually are, different. This somewhat simplifies the network planner's job. Now it is possible to overlay macro- and micro-cells even in mutually nonisolated areas. However, carrier frequency allocation becomes more difficult. Furthermore, it is difficult to overlay several micro-cells as carrier allocation and mutual interference avoidance becomes increasingly difficult. Consequently, here too, micro-cells are introduced for the initial purpose of filling in the gaps in network coverage.

4.1.3 Pico-cell

Pico-cells are yet smaller than micro-cells, although no standard definition exists. A pico-cell is generally in an isolated, confined area. This can be in an auditorium, inside a plane, or on a train platform. The capabilities of a pico-cell base station, its processing power, and antenna size are similar to, or less than, that of a micro-cell.

Pico-cells are designed to fill in gaps left by macro- and micro-cell structures. As they are generally well isolated, frequency planning is a significant issue: low transmit powers will further ensure minimal interference to its surroundings.

Connections to a central network can be via a macro- or a micro-cell base station. The connection may itself be wireless, and at times the pico-cell may act as a subscriber to the overlapping cell structure. Other wireless connections may be fixed wireless, or even satellite links for planes.

4.1.4 Umbrella structure

In practice, many macro-, micro-, and pico-cells coexist in order to cover an area. An umbrella structure can exist where coverage areas overlap, but mutual interference is avoided through frequency reuse, or heterogeneous technology utilisation, or mutual isolation, for example in indoor and outdoor topologies. We will discuss more about the umbrella structure in Section 7.1.2.

4.1.5 Repeaters

Another element in traditional network plans are repeaters as illustrated in Figure 4.7. These devices are simple; they consist generally of an outdoor receiving/transmitting antenna, an amplifier, and an indoor transmitting/receiving antenna. The function of a repeater is to increase the signal power within a confined area in the same way that a micro-cell or a pico-cell base station does. However, a repeater is simpler as it does not carry out any signal processing by itself. It only bridges a well-covered outdoor environment with a well-isolated indoor, often in an underground environment.

Figure 4.7 A repeater

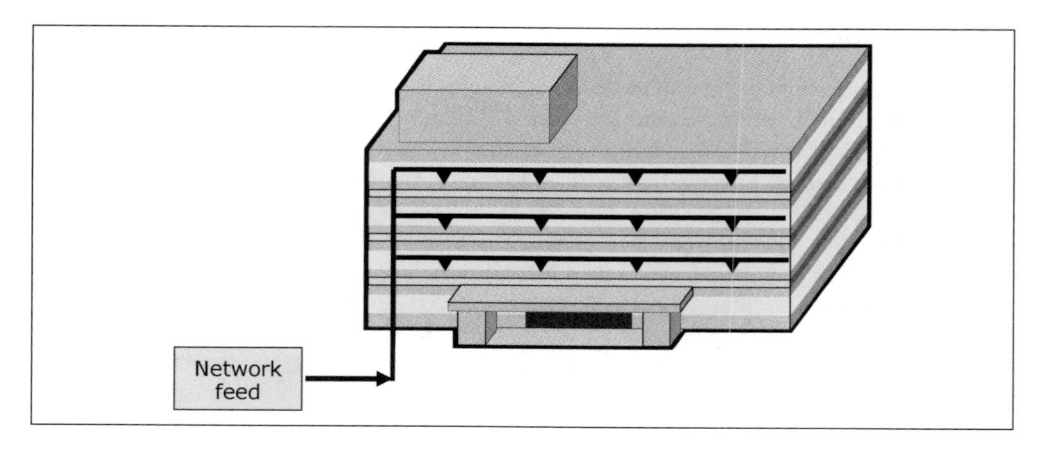

Figure 4.8 In-building coverage using distributed antenna systems

4.1.6 Distributed antenna systems

A solution for coverage inside a building is illustrated in Figure 4.8. Here, the radio signal is sent inside a building using fixed links, which are connected to antennas within the building. The signal level is kept high enough to provide needed SIR, while interference to the outside umbrella macro-cell remains small.

4.2 Wireless LAN Cellular Structure

As discussed in Chapter 2, IEEE 802.11-based Wireless LAN systems are a candidate for broadband systems. A possible cell structure for these systems is similar to the above macro-cell definition. Although several technologies have been developed to increase the coverage area of WLAN systems, due to the CSMA nature of their operation, coverage areas are likely to be small and comparable to a micro- or a pico-cell. The difference here is that there is no overlaying macro-cell structure to provide large area coverage.

Systems based on IEEE 802.11 technologies will, therefore, typically require a large number of cell sites to provide continuous coverage. This leads to large infrastructure cost, which means that such topologies are not economical. Present business models of public WLAN operators are mostly aimed at the so-called *hot-spot coverage*, such as within coffee shops, train stations, airports, and so on.

As illustrated in Figure 4.9, one solution to achieving seamless coverage is by combining hot spots with a public, wide-area coverage system in a hybrid topology. Here, a WLAN operator enters a business partnership with a public network operator. The end-user connects to the hot spot for higher (and usually less expensive) data-rate connectivity when within the hot-spot coverage, and to the wide-area network (WCDMA, GSM, or WiMAX) when outside the hot-spot area. In practice, cellular operators provide a WLAN service as a complementary service to their wide-area service. An example is DoCoMo's Mzone service as shown in Box 'DoCoMo Mzone'.

A further possibility exists in using multi-hop technology. Here, a subset of WLAN access points are connected to the internet. Those access points that are not connected,

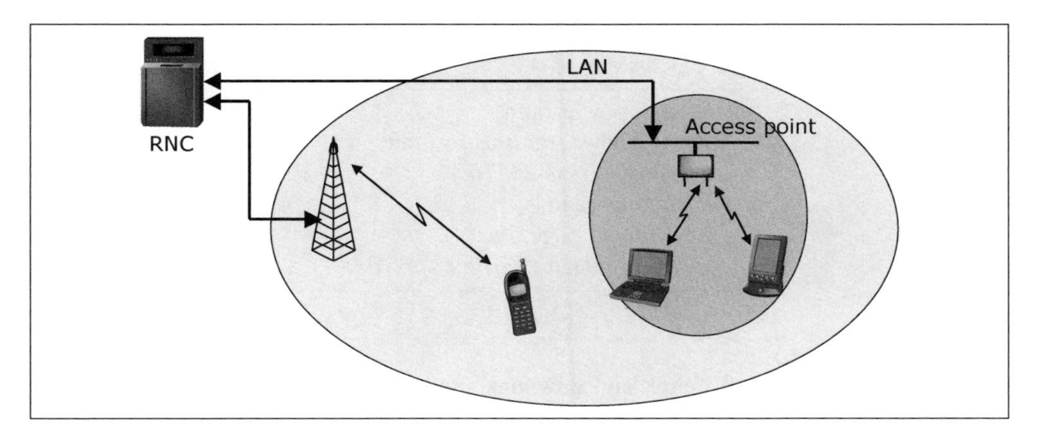

Figure 4.9 Hybrid networking using macro-cells and WLAN access points

DoCoMo Mzone

This is DoCoMo's Wireless LAN service and is available in many train stations, airports, and other public places. These areas are referred to as Mzone. DoCoMo also has a series of handsets, 900iL which are Dual WCDMA–WLAN capable. Similar services are offered by other operators in Japan and elsewhere.

transmit their signals to an access point that is connected. This can provide connectivity, but per-user transmission rates decrease quickly, as an access point's resources must now be shared among many more users.

4.3 Distributed Base Stations

We have already discussed coverage and how path loss varies as a function of the carrier frequency (Section 2.9). As discussed in Chapter 2, higher transmission rates require higher frequency bandwidths. As the lower bands are crowded, the new broadband wireless systems are expected to operate in higher frequency bands. Here, path loss is relatively higher as compared with lower bands as discussed in Section 2.9. Higher path loss, combined with higher transmission power requirements due to higher transmission rates, mean that very high transmit powers are required to ensure a desired level of signal quality. Table 4.1 exemplifies the ratio of required power for a 3G service of 2 Mbps operating in the 2 GHz band, and for a possible 4G service of 20 Mbps operating in the 5 GHz band. The 4G system requires an extra 14 dB, or 25 times the transmission power for similar BER reception quality.

While it may be possible for a base station to transmit at high powers, it is nearly impossible for a small, battery-powered user equipment to do the same. One solution is by making the uplink signal to reach the base station through multiple hops, that is, the uplink signal travels via several nodes, each a short distance requiring a small transmission power, to reach the central base station. This structure is called *multi-hop*. With fast transmission

Table 4.1 Extra required transmission power for beyond 3G systems

Extra required power ratio	
Due to higher transmission rate (20 Mbps versus 2 Mbps)	10 dB
Due to operating band	
Hata model page 59	
Due to operating band (5 GHz vs 2 GHz)	4 dB
Total	14 dB

rate requirements for both uplink and downlink, multi-hop methods have been proposed for both directions.

4.3.1 Uplink distributed base stations

An initial multi-hop structure is known as *distributed base station* and is aimed at solving the uplink transmission power shortfall. Several distributed base station cell structures have been proposed and researched in recent years. The initial design for a distributed base station is illustrated in Figure 4.10. Here, downlink transmissions are from a central base station. The smaller base stations distributed throughout the coverage area receive uplink signals and relay them to the central base station.

In practice, a distributed structure can easily be implemented. The smaller base station can be a simple receiver antenna which relays the received, unprocessed signal to the central station. The only concern is the availability of wired (or wireless) links connecting the outpost to the centre.

4.3.2 Downlink distributed base stations

Distributed base station concept may also be implemented in the downlink. However, while a distributed base station concept with its uplink- and downlink- independent transmission can easily be implemented, the same is not true if the smaller base stations were to transmit

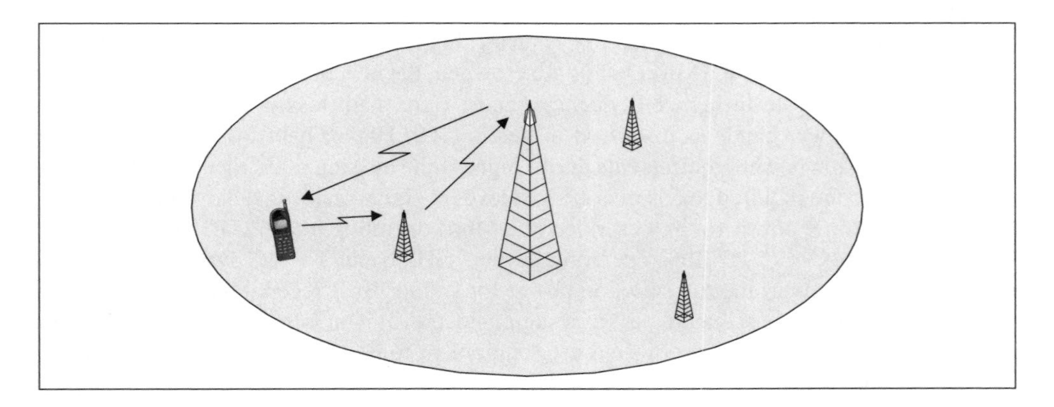

Figure 4.10 An uplink distributed base station

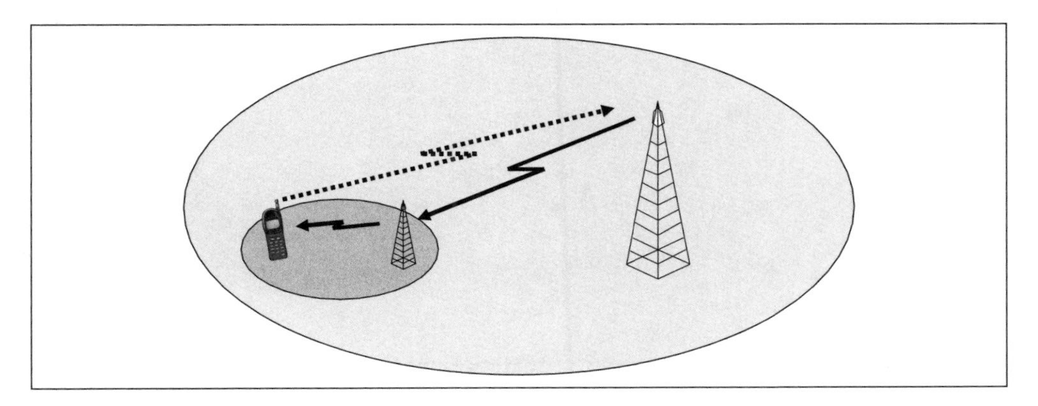

Figure 4.11 A downlink distributed base station

as well as receive. This causes interference overlap similar to the cases of macro- and microcell overlay. As a result, new frequency planning needs to be carried out to ensure that the central base station and relaying base station do not mutually interfere. The operation is illustrated in Figure 4.11. Frequency resource management for downlink–uplink distributed base station structure is an important research area, and is strongly affected by the choice of duplex modes, FDD, and TDD.

FDD structure

In one method for an FDD system, the distributed base station structure requires independent carrier frequencies for terminal uplink and initial downlink hops, as well as for each of the subsequent hops from relaying stations. This is illustrated in Figure 4.12. It is obviously quite difficult to manage frequency resource allocation for such a system. Furthermore, such a structure could lead to low frequency-utilisation efficiency.

A better solution is in using different time slots to transmit a signal at each hop. This leads to a TDD duplexing structure. While an overall FDD uplink and downlink may be maintained, each hop is processed in a TDD manner.

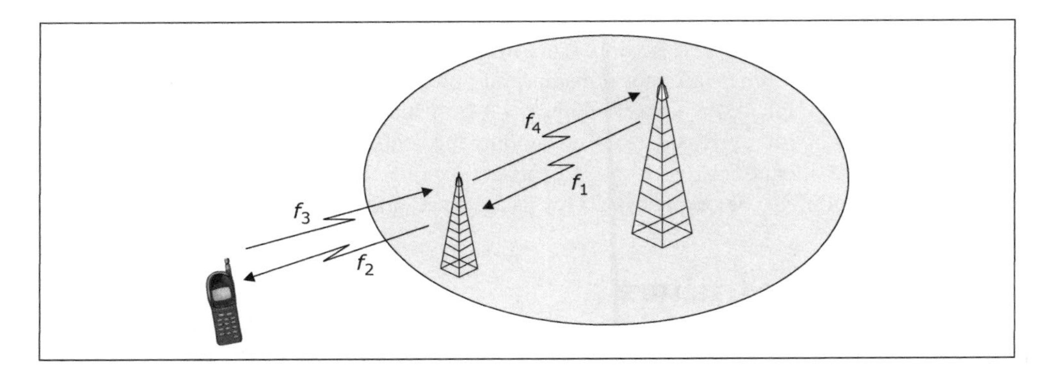

Figure 4.12 FDD-based frequency allocation for a distributed base station

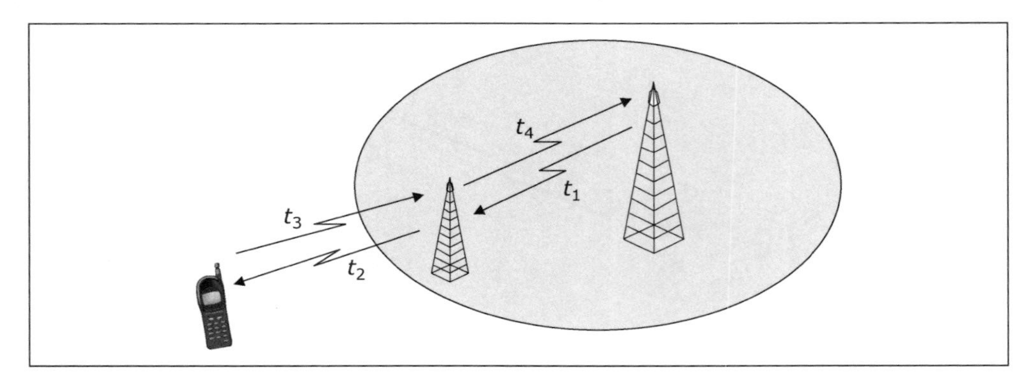

Figure 4.13 TDD-based signal transmission for a distributed base station

TDD structure

The TDD-based operation of a distributed base station systems is illustrated in Figure 4.13. A data packet travels from a base station to an end-user device over several hops. The central base station determines which slots are used for its uplink and downlink communications. The subsequent hops may be carried out using the same uplink and downlink slot structure, or by using a dynamic slot allocation strategy. The latter is more attractive as it realises a much higher frequency-utilisation efficiency. In this method, the relaying stations choose to send multi-hop packets onward, or receive them from mobiles in any slots. The only restriction is that the interference to other users of the slot is maintained below a design level. Again, this is a rather new area of research, which is expected to grow in importance as demand for higher transmission rates of the broadband wireless systems increases.

4.3.3 Public–private multi-hop

So far, we have defined multi-hop systems in terms of public infrastructure. Base stations and sub-base stations were used to transmit signals to, and receive signals from an end-user mobile (or fixed) station. Another topology uses end-user devices for multi-hop communications. The premise here is that a mobile signal may be relayed to another mobile who may then send it onto a central base station. This may be done because the relaying mobile is better situated, or that it has a higher transmission power capability. The same may be done for downlink communications as well. Figure 4.14 illustrates this concept. The research in this area is also at the early stages. Key among the technical issues to be resolved are efficient resource allocations, as discussed earlier. Among the business issues are how the relaying user is going to be compensated for its 'service' and how tariffs may be calculated.

4.4 Mini-cell Structure

We now return to the overlay concept discussed in Section 4.3. There, we maintained that overlay cellular structure is practical if, for example, macro- and micro-cell areas are mutually isolated, as in the case of a subway train station. Figure 4.6 showed how mutual

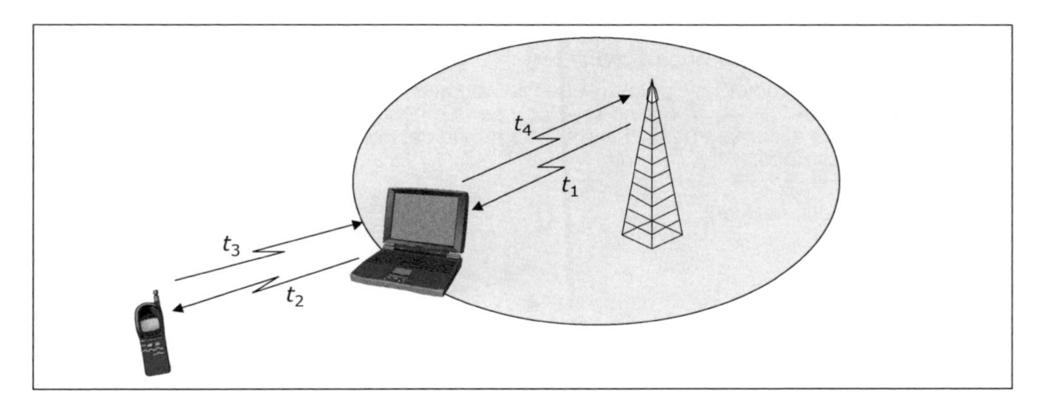

Figure 4.14 Multi-hop communications using end-user devices

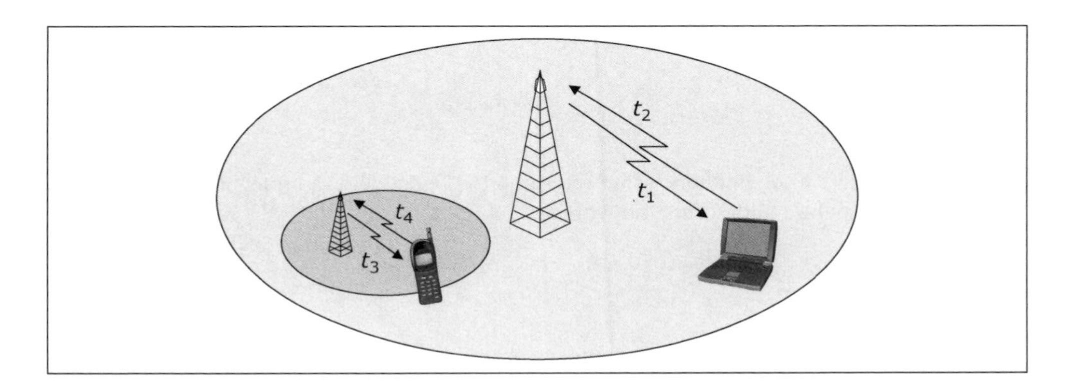

Figure 4.15 Cell overlay in TDD mini base-station structure

interference could result in areas with poor received SIR. We also stated that such an operation may be possible for FDD systems which use different carrier frequencies in each cell.

These conclusions are true, however, mainly for an FDD system. In a TDD system, it is possible to take advantage of the overlaid structure with dynamic resource allocation. The operation is illustrated in Figure 4.15. As different slots are used for uplink and downlink communications in the macro- and micro-cells, mutual interference characteristics are different from that of Figure 4.6.

This structure is similar to micro-cell base stations. And we refer to this structure as 'mini' base station to distinguish it from the macro- and micro-cell concepts discussed earlier. Mini base station can be very useful for filling in gaps, as well as for augmenting capacity requirements as demand rises. A TDD operator may first provide large area coverage using macro-cells. It can then provide high throughput in areas where there is demand by using one or more mini base station. The efficiency of a TDD system depends on how well dynamic slot allocation can be performed. This is another area of promising potential for research for broadband wireless systems. This approach can be utilised to construct a

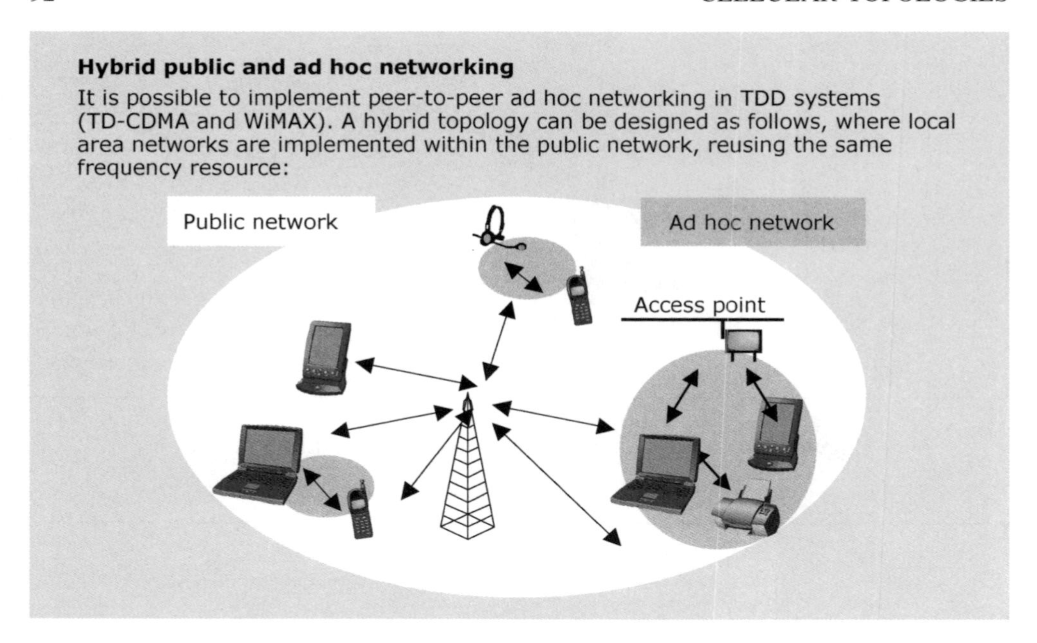

hybrid public and ad hoc network using the same frequency and radio technology as shown in Box 'Hybrid public and ad hoc networking'.

4.5 Handover

A mobile station travelling within a cellular communications system connects to the most suitable base station at any moment. Cellular communication systems are designed to facilitate handover of a user from one base station to another as dictated by channel conditions. Handover mechanisms have been specified for almost all cellular standards since the advent of analogue 1G systems.

Handover processes for circuit-switched voice communication systems of 2G and 3G are divided into two main categories: hard handover and soft handover. In hard handover, the transmission between an user equipment and a base station is stopped before transmissions to and from a new base station begin, that is, communications are carried out with one base station at a time. As illustrated in Figure 4.16, the MS stops its communications with BS1 before it starts communicating with BS2. In soft handover, a user equipment may be connected to more than one base station at a time. In this handover mode, data signals are transmitted from two or more base stations to a target mobile, which combines them to increase signal quality. In Figure 4.16, the MS continues to communicate with both BS1 and BS2 for some time, until the received signal from BS1 is lower than that of BS2 by more than a design margin. Soft handover techniques are used in CDMA voice systems, helping to improve connection quality at cell edge.

For the future 3G packet-switched systems and beyond (such as broadband wireless systems described in this text) hard handover process is expected to be used. This is mainly because of the large packet sizes, the use of HARQ techniques, and the difficulties in

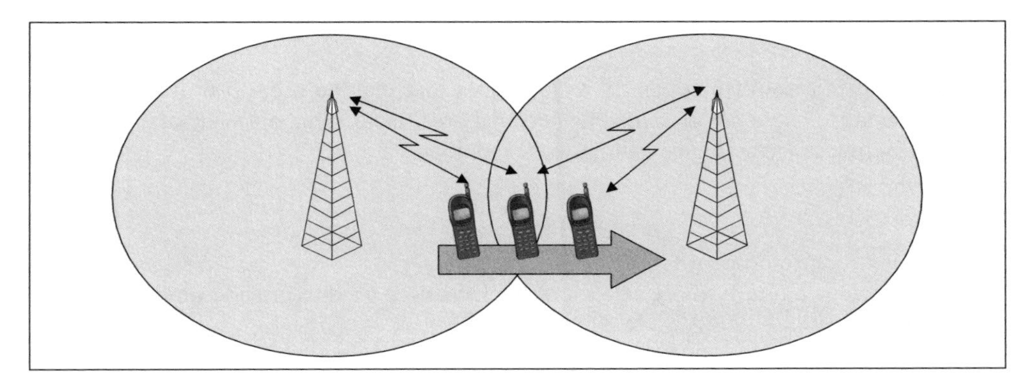

Figure 4.16 Hard and soft handover processes

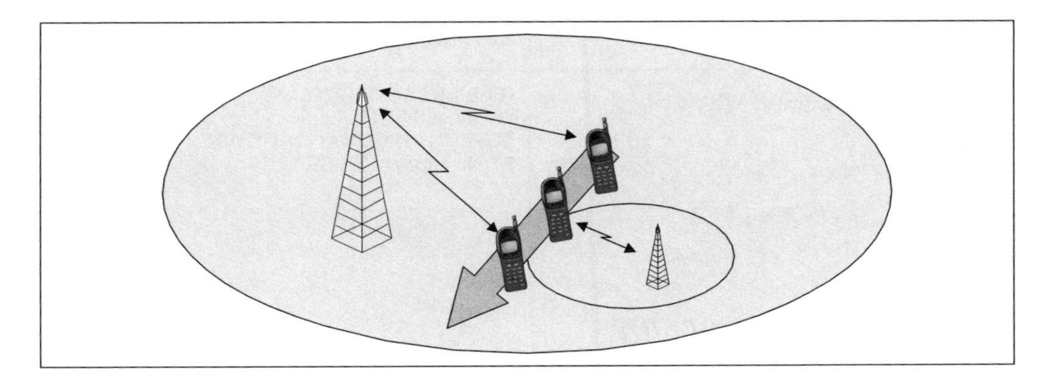

Figure 4.17 Hard handover in multi-structured cell topologies

managing multiple transmissions from multiple base stations. Hard handover in packet-based systems is a research topic, particularly within mobile IP services, destined to be carried over to the future broadband wireless systems.

As illustrated in Figure 4.17, a mobile station in an overlaid, multi-structured base-station system with macro, micro, and pico topologies, can connect to any of the base stations near it. The decision regarding which base station to use is based on the cost and rate of transmission as well as network traffic conditions. It is desirable that the most efficient connection is used at any time, and handovers are made as soon as required. However, the delays associated with handover processes make frequent handovers difficult. Furthermore, as packets are directed to and from a central network, as illustrated in Figure 4.17, only a limited number of base stations can be associated with a mobile station at any time. This is further restricted by the fact that a central network controller is connected to a limited number of base stations, and at times, inter-network handover is not readily possible. The smaller the cell size, the more frequently a mobile station crosses boundaries from one cell to another. This can result in handover failure, and dropped calls. This problem can occur in micro-cell based structures such as those encountered with Wireless LAN systems.

IEEE defines a set of hierarchical technologies that may be used to provide coverage over a wide area. These are classified as personal area, local area, and wide-area networks (see Box 'Wireless technologies and network topologies').

Wireless technologies and network topologies

IEEE envisions a future where several technologies are used to provide wireless connectivity over different networking topologies:

	Coverage	Technologies
Personal area network (PAN)	<10 m	IEEE 802.15: UWB, Bluetooth ETSI: HiperPAN
Local area network (LAN)	<100 m	IEEE 802.11: Wireless LAN ETSI: HiperLAN
Metropolitan/wide area network (MAN/WAN)	< 20 km	IEEE 802.20/802.16 Wireless MAN 3GPP(2): WCDMA/TD-CDMA ETSI: HiperACCESS

4.6 Ad hoc Networking

We end this chapter with a short discussion on ad hoc networks. Ad hoc networks can be defined as a group of devices that form communication links with each other on a temporary basis. A simple example is the link made between a headset and a mobile phone, where voice information is passed between the headset and the mobile, which in turn passes it on to the base station. A more complex example may be that of a meeting where files need to be shared between several notebook PCs. Another example would be a sensor network, where individual units/nodes communicate with each other to convey a set of information, relay the information forward to another node, and disband after successful completion of the information transfer. This is also known as peer-to-peer communication.

In general, networks with a base station/mobile station structure use a star topology, as shown in Figure 4.18, where many end-user equipment connect to a central processor. However, ad hoc networks follow a mesh topology, where a user can communicate information with several other users. This is illustrated in Figure 4.19. Although most future broadband wireless systems are expected to follow the traditional BS–MS star topology, some will include ad hoc modes to enable peer-to-peer communications.

Peer-to-peer can significantly add to the efficiency of a network. In this mode, two nearby users may exchange information without the involvement of a base station, and at very low transmit powers. As this reduces the load on the central network, and resource utilisation can be quite low, efficiency can increase significantly.

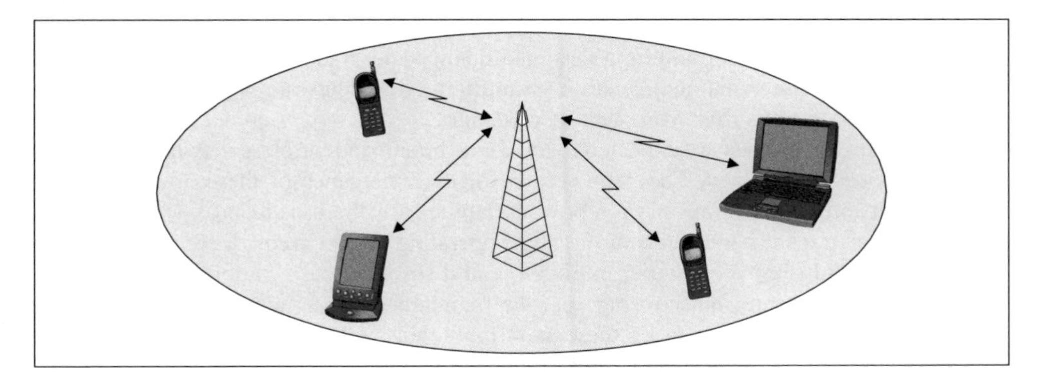

Figure 4.18 Star topology

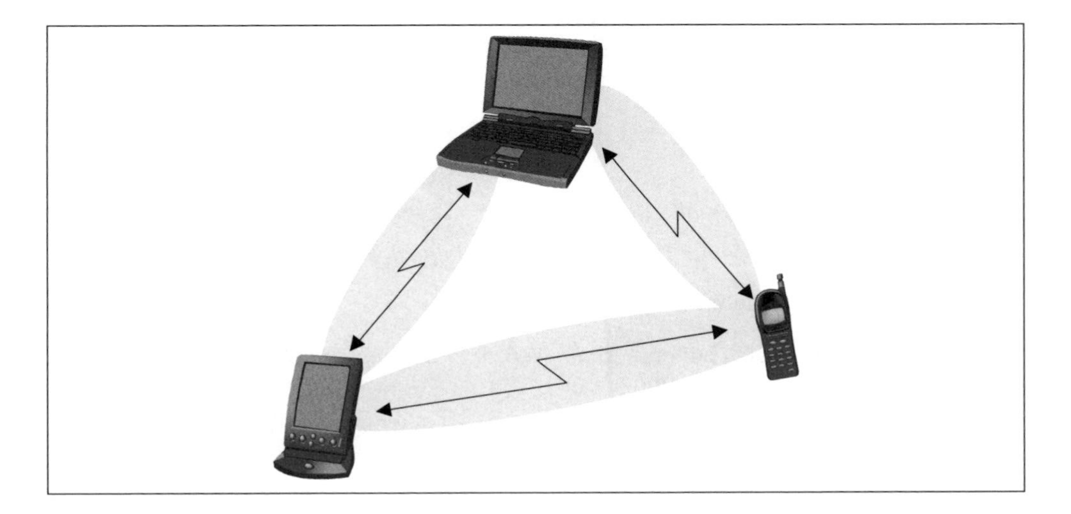

Figure 4.19 Mesh topology

Combined star-mesh networking can be implemented using heterogeneous technologies, such as a combination of Bluetooth and GSM technologies in the headset example above. It may also be implemented using a single technology, such as the ad hoc mode of IEEE 802.11 WLAN. Generally, TDD-based systems can easily accommodate joint ad hoc and public modes, and some research is ongoing on joint hybrid research in TD-CDMA and WiMAX systems.

Summary

This chapter has given a summary of network topologies for wireless communication systems. Many of these topologies have been used since the appearance of the first generation of mobile communications. They are designed to ensure that service coverage is maintained over a geographical area, providing a designed quality and grade of service. A mixture of these topologies ensures that areas with high traffic requirements are well served, and that populated areas do not suffer from lack of coverage.

Several new topologies are expected to appear as broadband wireless networks are rolled out and services are offered. The main reason for the emergence of these new topologies is the significantly higher transmission powers required for the broadband systems, as they promise higher transmission rates and as their operating carrier frequencies are shifted to higher bands. Multi-hop techniques, mini-cell, and distributed base station structures, have been proposed to ensure that coverage can be maintained throughout a service area. We discussed how the introduction of these new topologies will affect network design and operation, specially in areas such as those requiring handover.

We also discussed ad hoc, and peer-to-peer communications. Ad hoc networking has generally been offered using heterogeneous technologies, but some broadband access technologies promise joint public and ad hoc modes. Such modes of operation are expected to significantly increase frequency-utilisation efficiency.

Further Reading

- On Mobile IP:
  - Perkins, C. E., *Mobile IP Design Principles and Practices*, Prentice Hall, 1998.
  - Solomon, J., *Mobile IP the Internet Unplugged*, Prentice Hall, 1998.
- On Resource allocation:
  - Laiho, J., et al., (Editors), *Radio Network Planning and Optimisation for UMTS*, John Wiley & Sons, 2002.
  - Zander, J., *Radio Resource Management for Wireless Networks*, Artech House Publishers, 2001.
- On Wireless World Research Forum:
  - Tafazolli, R., (Editor), *Technologies for the Wireless Future: Wireless World Research Forum (WWRF)*, John Wiley & Sons, 2004.
  - WWRF website `http://www.wireless-world-research.org/` 2005.

5

Cost of Spectrum

The most basic resource in wireless communications is the frequency spectrum over which radio transmissions are made. Efficient utilisation of this resource is important considering the cost of building a total system. The spectrum is a national resource: it is owned and managed by a national government. Public telecommunications are only one of the many services that use radio spectrum: other major uses include TV broadcasting services, emergency services, satellite operations, military telecommunications, and so forth. An example of spectrum allocations in the 1.7 ~ 3 GHz band in Japan is shown in Figure 5.1. This is a typical allocation, and more or less similar bands are allocated for similar services elsewhere in the world.

Some parts of the spectrum are licensed to an operator for a definite period of time, at a certain charge and with certain conditions. The license is usually renewed as long as there is a need for the operator's services within the government's telecommunications policy. However, this is not automatic and technically a government may terminate an operators' lease. At times, an operator would have to pay large fees for the right to use a spectrum. Box 'How much was paid' is extracted from ITU figures, and shows how much operators paid in several European countries for 3G licenses. Such a right is known as a *spectrum license*, and the corresponding spectrum is known as a *licensed band*.

There is another kind of spectrum, known as an *unlicensed band*. Such a spectrum may be used by operators and private users without the need for a license. The government retains ownership, but allows free use of the band with specific conditions such as transmission power limits, and so on.

As the next 10 years are expected to witness immense growth in broadband wireless services, a number of countries and regions have planned, and in some instances already allocated, frequency spectrum for these services. Table 5.1 lists the spectrum allocation, or plans to that effect, in some major regions. Sub 1 GHz band spectrum, is already used for 2G (GSM, IS-95) systems. The spectrum used for analogue 1G systems, such as 450-MHz band, as well as broadcast spectrum are being considered for 4G wireless DSL-type services.

Broadband Wireless Communications Business: An Introduction to the Costs and Benefits of New Technologies Riaz Esmailzadeh

How much was paid?

Many operators in Europe paid large amounts of money for 3G spectrum:

Country	Number of Licenses	Total paid ($M)	Paid per population ($)
Denmark	4	393	73
France	2	7622	130
Germany	6	38 860	473
Great Britain	5	29 466	495
Italy	5	9356	162
Sweden	4	None –"Beauty Contest"	

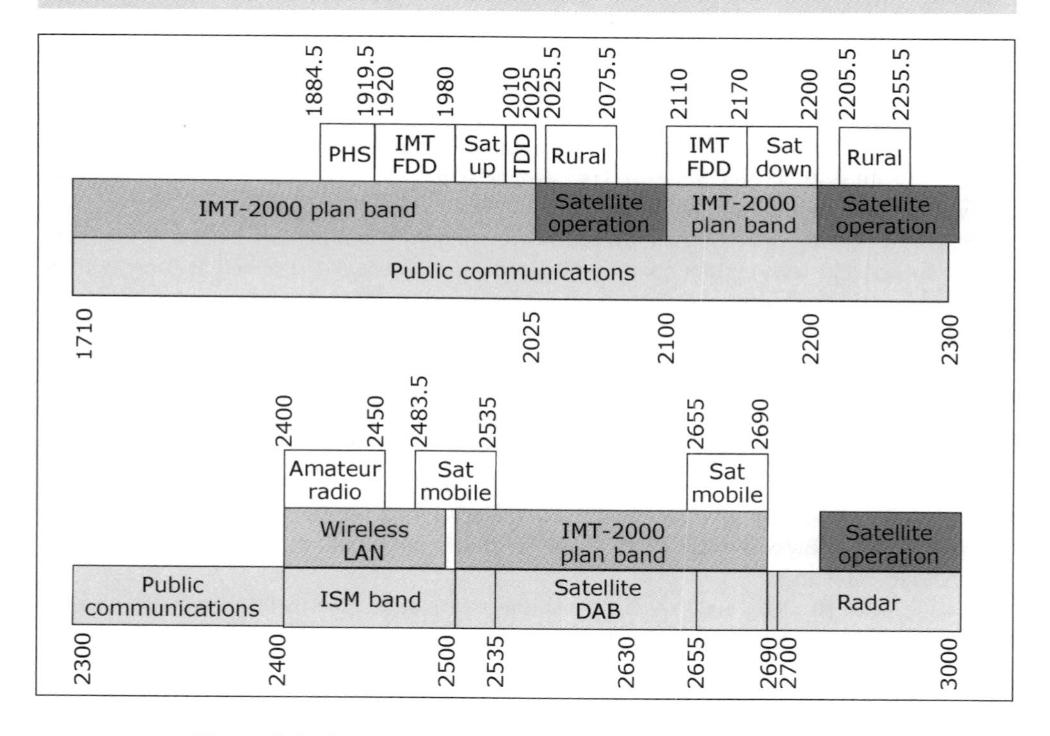

Figure 5.1 Spectrum allocations in the 2-GHz band in Japan

This chapter focuses on the costs associated with the use of a frequency band. Most of our discussion is on the frequency-utilisation efficiency of different technologies, but we will also address issues such as the spectrum required for delivering a certain quality of service (QoS) to a target market size. Roaming requirements and national harmonisation in

Table 5.1 Spectrum allocation overview for several major regions (B3G is beyond-3G, and FWA is fixed wireless access)

Region (GHz)	1.7–1.8 (GHz)	1.9–2.1 (GHz)	2.3 (GHz)	2.4 (GHz)	2.5 (GHz)	3.5 (GHz)	5.1–5.7 (GHz)	5.8 (GHz)
North America	2G/3G	2G/3G	B3G	WLAN	B3G	FWA	B3G/FWA	WLAN/B3G
Europe	2G/3G	2G/3G	B3G	WLAN	B3G	FWA	B3G/FWA	WLAN/B3G
Japan	2G/3G	2G/3G	B3G	WLAN	B3G	FWA	B3G/FWA	WLAN/B3G
China	2G/3G	2G/3G	B3G	WLAN	B3G	FWA	B3G/FWA	WLAN/B3G

spectrum allocation for broadband wireless services will also be discussed. These are all concerned with operations in the licensed band; however, we will also briefly discuss the frequency efficiency of services in the unlicensed band.

As discussed in Section 2.8, frequency-utilisation efficiency (as derived from system capacity) is a measure of the amount of information that can be carried by a communications system. For voice channels, this is the number of possible simultaneous voice calls. For data communication services, this is the average number of information bits that are correctly transmitted per second over a service area. Here, frequency-utilisation efficiency is also known as *normalised average throughput*. Other parameters such as peak throughput, or the total number of bits transmitted when channel conditions are optimal, and minimal possible throughput to a certain percentage of users (such as 95%) are also important indicators of a system's performance.

5.1 Voice Systems

We will first discuss about frequency-utilisation efficiency in terms of conventional voice-centric services in terms of voice channel per MHz of bandwidth. The total user capacity of a traditional mobile phone communications system may be defined as the total number of users that can simultaneously connect to the network, and carry on a conversation via the base station of a single cell. Because the total spectrum is limited, the capacity is limited. To deliver and maintain a certain QoS, the received signal power must be large in comparison with the receiver noise level as well as the interference from other devices that use the same frequency band (co-channel interference). As discussed in Chapter 2, mobile communication systems are generally defined to be noise limited or interference limited. In noise-limited systems, the signal-to-noise ratio (SNR) is significantly smaller than the signal to co-channel interference ratio (SIR). For the interference-limited systems, the reverse holds true. The first-generation FDMA systems are considered to be noise limited when their frequency reuse factor is sufficiently high.

5.1.1 FDMA systems

Calculation of voice channels for a 1G FDMA system is rather straightforward. The total bandwidth is divided by the frequency reuse factor to give the available bandwidth for one cell. This bandwidth is then divided by the bandwidth required for one voice channel to obtain the total number of available voice channels in the cell. Assuming a uniform cell structure, this number is the total voice channels that can be supported within a limited area, and is the frequency-utilisation efficiency when expressed in number of voice channels per MHz of bandwidth. See the example for the North American advanced mobile phone service (AMPS) in Box 'Voice channels in the AMPS standard'.

Voice channels in the AMPS standard

An example of an FDMA system is the AMPS standard. In AMPS, an operator's available bandwidth is divided into several parts for frequency reuse, and then each part is subdivided into frequency bands of 30 kHz, each of which is then used for one connection between a mobile user and a BS. With the system parameters as in the table, the maximum number of users that can be simultaneously connected to an AMPS system in each cell is calculated to be 23. AMPS system's frequency-utilisation efficiency in terms of voice channels per MHz of bandwidth is equal to 4.6.

Total available bandwidth	10 MHz (5 + 5 up and down)
Reuse factor	7–12
Single channel bandwidth	30 kHz
Voice channels in each cell	23

5.1.2 TDMA systems

As described in Chapter 1, in TDMA systems, several users access the central BS using the same frequency band, but at different times. They take turns connecting to the base station one at a time as shown in Figure 1.12. This technology is used in most 2G systems such as digital advanced mobile phone service (D-AMPS), PDC, and GSM. The use of digital technology allows the effective required bandwidth for one voice call to be substantially reduced. This increases the frequency-utilisation efficiency several fold as compared to FDMA. The D-AMPS system uses the same 30-kHz spectrum of the AMPS standard. But, this spectrum is now used to support three simultaneous voice calls instead of one. With the same frequency reuse factor, TDMA systems are also noise limited. However, as they use digital modulation, they can operate at much lower SNR levels. This is important in that it reduces the required transmission power of the mobile station (and base station), resulting in smaller batteries and longer talk times.

For an example of frequency-utilisation efficiency of TDMA systems, see Box 'Voice channels in the GSM standard'.

Voice channels in the GSM standard

An example of a TDMA system is the GSM standard. In GSM, a 200-kHz bandwidth is used to accommodate 8 voice channels. With system parameters as in the table, the maximum number of users that can be simultaneously connected to a GSM system in each cell is calculated to be 66. The frequency-utilisation efficiency is equal to 13.2 voice channels per MHz of bandwidth.

Total available bandwidth	10 MHz (5 + 5 up and down)
Reuse factor	3–4
Single carrier bandwidth	200 kHz
Voice channels per carrier	8

5.1.3 CDMA systems

As described in Section 1.8, CDMA systems work on the principle that two or more user signals may be transmitted, and distinctly received, in the same band as long as they use different codes for spreading their signals. After spreading, all signals occupy the same bandwidth and appear to each other as noise. At the receiver side, a de-spreading function is carried out that restores only the desired signal to the original narrow band.

The initial 2G IS-95 standard was used primarily for voice communications. The total voice channels can again be calculated much in the same way as in FDMA and TDMA systems. However, here we need to use two new parameters. First, owing to the fact that the same frequency is used throughout the system, the frequency reuse factor is equal to one. Within a cell, a certain percentage (up to 80%) of the codes may be used for voice traffic. Unfortunately, the co-channel interference from other users within neighbouring cells reduces this number to less than 50%. This means that the frequency reuse factor is *effectively* larger than one. Typical values are between 2 and 3. Using voice activation, the device is active only about 50% of the time. This reduces the total interference, and therefore reduces the effective reuse factor by half. Using this effective reuse factor, total system bandwidth, and the bandwidth required for a single voice channel, the frequency-utilisation efficiency can be calculated. For an example of CDMA, see Box 'Voice channels in the IS-95 standard'.

Voice channels in the IS-95 standard

An example of a CDMA system is the IS-95 standard. In IS-95, the carrier bandwidth is typically 1.6 MHz, which allows three carriers within a 5-MHz bandwidth. Spread spectrum technology is used to transmit many voice signals simultaneously. With system parameters as in the table, the maximum number of users that can be simultaneously connected to an IS-95 system in each cell is calculated to be about 120, allowing for voice activity. The frequency-utilisation efficiency is equal to 24 voice channels per MHz of bandwidth.

Total available bandwidth	10 MHz (5 + 5 up and down)
Carrier bandwidth	1.6 MHz
Reuse factor	1
Voice channels per carrier	~20
Voice activity factor	0.5

5.2 Data Systems

Although voice transmission rates are not equal for all digital systems, they are broadly equal. Voice encoders, or vocoders (devices that transfer an analogue voice signal to a digitised voice stream), offer a range of rates between just a few kbps up to 64 kbps or more. However, vocoders of most 2G and 3G systems support rates between 4 to 12 kbps, which, after channel coding, results in a 24- to 32-kbps signal. Therefore, the voice channel

Table 5.2 Data Communications services and requirements

Services	Required transmission rates	Delay sensitivity
File transfer	>1 Mbps	Low
SMS	<10 kbps	Moderate
MMS	Tens of kbps	Moderate
Video streaming	1 Mbps	Moderate
Audio streaming	100 kbps	Moderate
Video telephony	Hundreds of kbps	High

per MHz of bandwidth is a reasonable comparison of the frequency-utilisation efficiency of different systems. For data systems, however, this is not the case anymore. Indeed, the term 'data communications' refers to a collection of different services, each with a different transmission rate and delay requirements. Table 5.2 shows a list of data services that can be expected in the next generation of mobile communications systems.

We have already discussed that transmission rates in DSL systems depend on the distance between an exchange and a subscriber (Figure 1.2). Wireless transmission rates are also dependant on the distance between a base station and a user equipment. As calculated from the Shannon theorem (Section 2.8.1), the throughput is dependent on the received SNR. The received signal power level itself is dependent on path loss, which is a function of the distance between the transmitter and the receiver. We further discussed in Section 3.2 that as mobile communications systems are mostly interference limited, a better approximation for the Shannon capacity is to replace SNR ratio ($\frac{S}{N}$) by SINR ratio ($\frac{S}{I+N}$).

Figure 5.2 illustrates the SINR for a cellular system with frequency reuse factor of 1. It can be seen that the SINR deteriorates as the distance between a base station and an

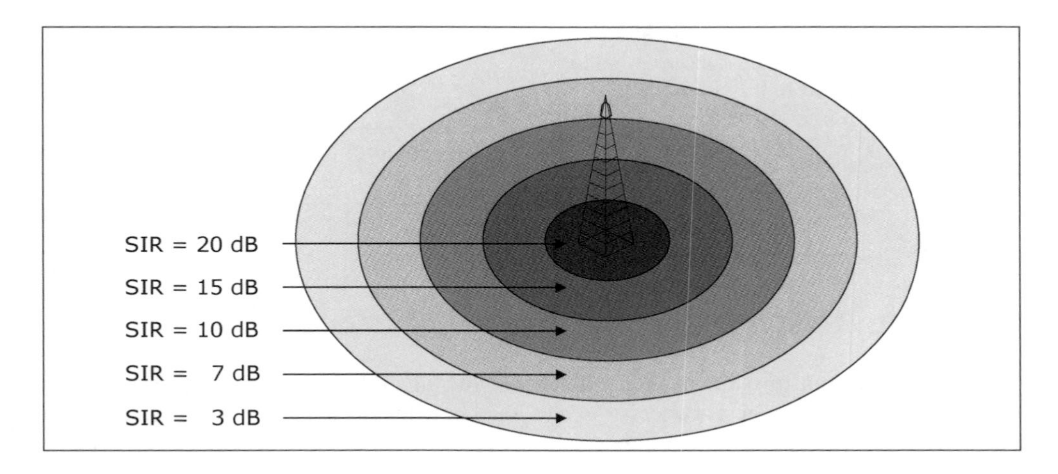

Figure 5.2 Signal-to-interference ratios in cellular systems

end-user equipment increases, resulting in reduced throughput values for different end-user equipment within the cell, as illustrated by Figure 3.2. The reason is that, in the downlink, interference sources are other base stations, and naturally, as an end-user equipment moves away from the base station, the received power decreases and the interference level increases. The shape and values in the figure are for illustrative purposes. In practice, physical objects in a cell result in a highly complex SINR distribution.

While voice users require uniform transmission rates regardless of where they are located within a cell, data users may tolerate different transmission rates, and an operator may design the network similar to a DSL model, where users nearer to the base station enjoy a better QoS.

Therefore, a better measure of frequency-utilisation efficiency in data systems with variable transmission rates is 'throughput', which is defined by the Shannon theorem as the number of information bits that are successfully transmitted per second. Again, since the throughput figure is a variable, depending on the location of a particular end-user equipment, there are several throughput figures of interest.

5.2.1 Peak throughput

Peak throughput refers to the highest possible transmission rate available within the system. It is obtained when SINR is maximal, and translates into the highest possible throughput based on the Shannon theorem. High throughput may be obtained by using higher orders of modulation and low coding. As discussed in Section 2.2, higher orders of modulation can increase the throughput per transmitted symbol. Furthermore, lower coding rates, as discussed in Section 2.6.3, also increase the transmission efficiency. All of these techniques can be utilised if a higher SINR is available.

We have discussed about adaptive coding and modulation (ACM), which is used to maximise the frequency-utilisation efficiency, in Section 3.2. In this method, the channel conditions are regularly monitored by a transmitter in order to find the SINR at an intended receiver. On the basis of this SINR value, the most suitable combination of channel coding and signal modulations is chosen to facilitate high throughput. Typical SINR levels necessary for several different combinations of modulation and coding rates in fading channel environment are listed in Table 3.1. The peak throughput is clearly obtained when the highest ACM level is used and when SINR is very high.

Peak throughput values are quite similar for different broadband wireless access technologies. These figures are dependant upon the bandwidth of the system, which, in turn, determines the maximum symbol rates, and the highest supported modulation rate. What mainly distinguishes different broadband wireless access technologies is the way they deal with co-channel interference. Since peak throughput is available only in areas where interference power levels are low, most of these technologies have similar throughput figures relative to their bandwidths.

5.2.2 Average throughput

Because peak throughput is available only in a small percentage of a base station coverage area, a better measure is average throughput, which indicates how a base station delivers service on average to the generality of users. This means (referring to Figure 5.2) the weighting of possible transmission rates of Table 5.2 is according to the probability that a user within the cell experiences an SINR level within a certain range, which in turn corresponds to a certain ACM level.

While peak throughput values are quite similar for different broadband wireless technologies, it is the average throughput figures that can be quite different. We shall discuss the different peak and throughput figures for several broadband wireless technologies in the following text.

The peak and average throughput figures when divided by the system bandwidth yield frequency-utilisation efficiency in bits per second per hertz.

5.2.3 Minimum throughput

Another useful measure of throughput is what a user experiences at the cell edge. Of course, the absolute minimum throughput will always remain zero (it is impractical to provide 100% coverage, and the areas that are not covered will naturally have zero throughput). However, a measure of how low a throughput may be before communications are discontinued can be referred to as minimum throughput. As can be expected, minimum throughput is related to the lowest possible ACM combination as well as to the maximum possible spreading factor when signal spreading is used (as in CDMA-based systems).

The minimum throughput figure can be important to those operators who need to guarantee a certain grade of service. When a connection is available, the minimum possible transmission rate could be important to customers. We calculate this figure only for WCDMA systems, as an example.

5.3 Data Throughput Efficiency

There exist several candidate technologies for broadband wireless services. One group is based on the present 3G standards, such as WCDMA, CDMA-2000, and TD-CDMA. We will deal with two specific standards, one that is based on 3G WCDMA high-speed downlink shared packet access (HSDPA), and the other that is based on 3G TD-CDMA HSDPA specifications. Other technologies have been newly developed and are yet to be standardised. Two of them are included: one is the WiMAX standard currently under development by IEEE. The other is a system by NTT DoCoMo, Inc, a method that is being proposed to enhance 3G HSDPA standards. Because this uses a very wide bandwidth, it provides the highest throughput among the methods discussed here. As can be seen from the table, these systems use a variety of different access technologies: TDMA, CDMA, and OFDM. Furthermore, they have very different total bandwidths, from 5 to 140 MHz. Thus, the available downlink and uplink throughput figures vary broadly from system to system. Therefore, it is the frequency-utilisation efficiency that is of greater interest. Table 5.3 summarises the parameters of the above mentioned technologies.

5.3.1 WCDMA HSDPA

HSDPA is a part of the 3GPP WCDMA standards. It is specified to facilitate high-speed and efficient downlink packet-switched communications. Table 5.4 summarises the major HSDPA parameters. The same bandwidth and chip rate as were used for earlier releases

Table 5.3 Broadband wireless access technologies

Technology	Duplex method	Downlink access	Uplink access	Downlink bandwidth	Uplink bandwidth
HSDPA	FDD	CDMA+TDMA	CDMA	5 MHz	5 MHz
TD-CDMA	TDD	CDMA+TDMA		5, 10 MHz	
WiMAX	FDD/TDD	OFDM+TDMA		1.25–20 MHz	
DoCoMo	FDD	OFDM+TDMA+CDMA	CDMA+TDMA	100 MHz	40 MHz

Table 5.4 WCDMA HSDPA parameters

Bandwidth	5 MHz
Chip rate	3.84 Mcps
Modulation	QPSK and 16-QAM
Channel coding	Turbo code
Effective coding rates	1/4, 1/2, and 3/4
Spreading factor	16
Multi-code operation	Up to 15 codes

of WCDMA are specified. HSDPA uses the ACM technique to take advantage of higher possible transmission rates when channel conditions are good and to remove the requirement for precise power control.

Peak throughput

The theoretical peak throughput of a WCDMA HSDPA system is easy to calculate. It is directly proportional to the highest modulation rate (16-QAM) and the highest coding rate. Hybrid ARQ techniques, specifically incremental redundancy, can bring the channel coding rate as close to one as possible. In the calculation in the following text, we use rate 3/4. The peak rate is also proportional to system bandwidth, which in this case is the utilised bandwidth in the form of downlink chip rate of 3.84 mega chip per second (Mcps). The total throughput is then the product of symbol rate, which at its highest is equal to chip rate, data bits per symbol as listed in Table 2.1, channel coding rate and information bit to data bit ratio. See Box 'Peak throughput' for the calculation of peak throughput in a WCDMA system.

The theoretical peak throughput is difficult to achieve except for a very limited portion of cell coverage area. This is because the required SINR as shown in Table 3.1 is achievable only in the areas near the base station.

Average throughput

The average SINR can be calculated for different areas within a cell as illustrated in Figure 5.2. This average SINR combined with values from Table 3.1 yield the average throughput for a WCDMA HSDPA system. SINR in a CDMA system can be defined as follows. It is the ratio of the transmitted signal power S to the sum of received noise power

Peak throughput

An example for a WCDMA system is calculated below. It should be noted that peak throughput figures of up to 14 Mbps have been reported. These require higher coding rates and better information efficiency than listed in this example.

Chip rate	3.84 Mcps
Spreading factor (chip per symbol)	1
Data bits per symbol (16 QAM)	4
Channel coding rate	3/4
Information bits/transmitted bits	95%
Peak throughput	10.944 Mbps

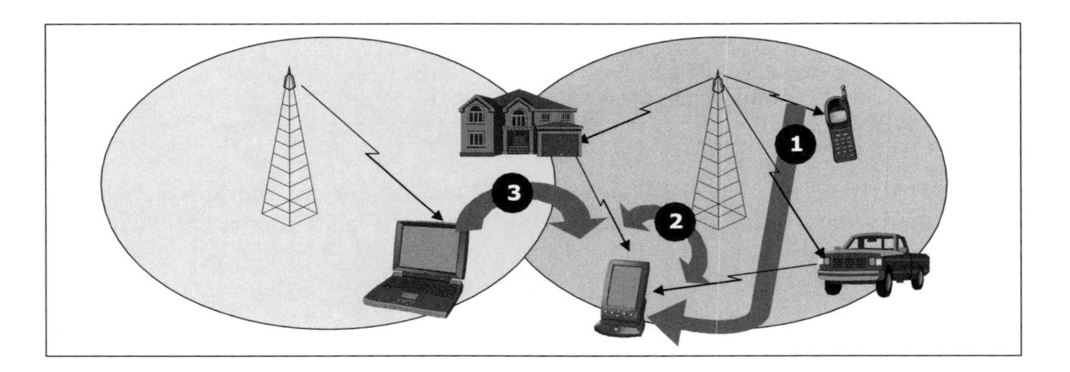

Figure 5.3 Sources of interference in a CDMA system

N and interference power I. While S can be controlled and N can be ascertained, I is a function of the transmission channel environment. It has been shown in Figure 5.2 how SINR decreases as the distance between a base station and a user equipment increases. Figure 5.3 shows the sources of interference for a received signal. Although some interference is indeed from neighbouring cells, the source of other interferences is from its own cell:

$$\text{SINR} = \frac{S}{N + I_{\text{other}} + I_{\text{own}}}$$

Own-cell interference is partly due to downlink control signalling and partly due to multipath as shown in Figure 5.3. The former can be minimised through joint detection, and the latter can be minimised through channel equalisation , as discussed in Section 2.3. However, SINR will be limited to a certain level, and on the basis of this an ACM combination can be used.

It should be noted that average SINR values depend upon channel conditions as well as on how well joint detection and channel equalisation can be performed. A typical calculation for an HSDPA average throughput is shown in Box 'Average throughput'.

Average throughput

Assuming the following SIR distribution table, the average throughput for a WCDMA system is calculated by summing up the SIR densities multiplied by their corresponding throughput.

SIR (dB)	Density (%)	Throughput
12	10	12 Mbps
9	25	5 Mbps
6	35	3 Mbps
3	25	1 Mbps
0	5	30 kbps
	Total	3.75 Mbps

Minimum throughput

The minimum throughput is calculated similar to the maximum throughput. The symbol rate may be quite low, as high spreading factors can be used. A typical minimum throughput may be as low as a few tens of kbps. See Box 'Minimum throughput'.

Minimum throughput

An example for the WCDMA system is calculated below.

Chip rate	3.84 Mcps
Spreading factor (chip/symbol)	256
Data bits per symbol (QPSK)	2
Channel coding rate	1/3
Information bits/transmitted bits	95%
Minimum throughput	9.5 kbps

5.3.2 WCDMA uplink

The uplink throughput of a CDMA system is calculated in a similar way to the downlink. While 3GPP is specifying HSUPA standard which uses similar techniques as HSDPA, we use the available figures from release six of the standard. This uses QPSK modulation only and variable channel coding. Typical throughput figures as well as frequency utilisation efficiency figures for both WCDMA downlink and uplink are listed in Table 5.5. Figures for the uplink are generally lower owing to the use of lower modulation orders.

5.3.3 TD-CDMA HSDPA

TD-CDMA system parameters are shown in Table 5.6. The throughput calculations are similar to those of WCDMA given earlier. Two major differences between TD-CDMA and

Table 5.5 Typical downlink and uplink throughput figures for WCDMA

	Downlink		Uplink	
	(Mbps)	(bps/Hz)	(Mbps)	(bps/Hz)
Peak	10.8	2.16	2.8	0.56
Average	1.38	0.28	0.8	0.16

Table 5.6 TD-CDMA parameters

Bandwidth	5 and 10 MHz
Chip rate	3.84 and 7.68 Mcps
Modulation	QPSK and 16-QAM
Channel coding	Convolutional and turbo
Effective coding rates	1/4, 1/2, and 3/4
Spreading factor	1–16
Multi-code operation	Up to 16 codes

WCDMA are (1) the use of TDD, and (2) extensive use of joint detection in both base station and end-user equipment. The use of TDD provides flexibility in the allocation of resources to either of downlink and uplink as traffic requirements change. Joint detection can be implemented more easily in TD-CDMA owing to its TDMA features and lower spreading codes. Furthermore, the new development of joint detection allows for reductions in interference from other cells (I_{other}) in the equation on page 106). This means that although peak throughput figures are similar for WCDMA and TD-CDMA, average throughput figures are different. Furthermore, a 10-MHz option has been specified for TD-CDMA, and therefore, peak throughput figures can be twice that of WCDMA. Throughput figures are summarised in Table 5.7. TD-CDMA is relatively unknown compared with WCDMA, and a list of the players in the field is shown in Box 'TD-CDMA players'.

5.3.4 WiMAX

As discussed in the previous section, both peak and average throughput figures are affected by co-channel interference. Of these, the own-cell interference I_{own} is caused partly by multi-path propagation in the transmission channel. As discussed in Section 2.3, multi-path

Table 5.7 Throughput figures for TD-CDMA downlink and uplink, for a 3.84 Mcps system, 12 traffic + 3 control slots in each frame

	Downlink normalised throughput		Uplink normalised throughput	
	(Mbps)	(bit/s/Hz)	(Mbps)	(bit/s/Hz)
Peak	8.0	1.6	3.4	0.7
Average	2.5	0.5	1.5	0.3

TD-CDMA players

TD-CDMA technology has been standardised through 3GPP. UMTS TDD alliance is an association of the TD-CDMA players: developers, manufacturers, operators, and academia.
IPWireless: has been the main developer of network equipment and chipsets.
Manufacturers: A number of manufacturers are entering the field. One major manufacturer in UTStarcom.
Operators: Many operators in the world have started commercial services using the standard and many are conducting trials.
T-Mobile in the Czech Republic has announced they will deploy a nation-wide network. In Japan **IPMobile** is in the process of obtaining license to operate a TD-CDMA network.

propagation leads to frequency-selective fading. While CDMA systems take advantage of multi-path signal arrival for diversity combining, one disadvantage remains. The signals of each path add to inter-path and inter-symbol interference, raising the level of I_{own}.

OFDM systems have an advantage over CDMA systems in that they do not suffer from frequency-selective fading. OFDM sub-carriers are, by design, narrow enough so that the fading remains flat within the sub-carrier. Furthermore, adding a guard-time interval between transmitted symbols deals with the multi-path signal arrival and reduces inter-symbol interference. These in turn reduce I_{own}. Therefore, the only major source of remaining interference is inter-cell interference, as illustrated in Figure 5.4. However, on the basis of channel conditions (such as a long delay profile), inter-symbol interference may exist. Inter-cell interference can lead to large performance degradation. One solution is to

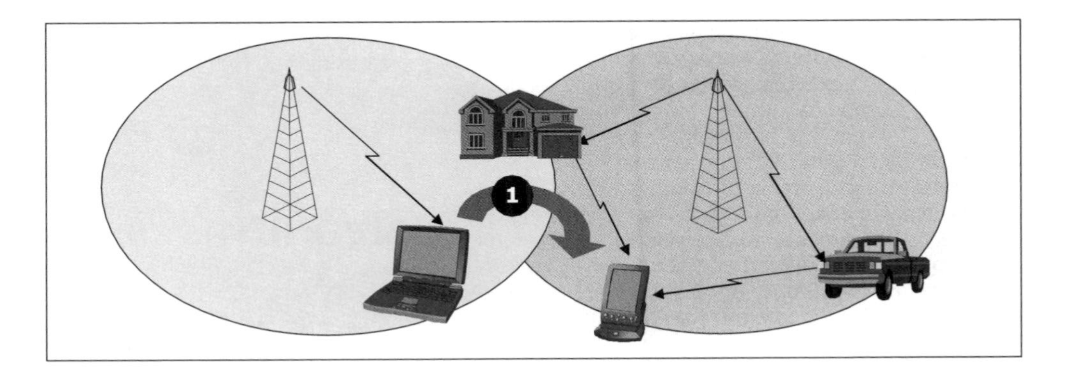

Figure 5.4 Sources of interference in an OFDM system

set the frequency reuse factors to three (as opposed to one for CDMA systems). This, in turn, reduces frequency-utilisation efficiency.

A standardisation process for broadband wireless systems parallel to that of 3GPP takes place within the IEEE 802 standard body. In particular, two sub-groups, 802.16 and 802.20, have developed several specifications for mobile broadband communications. Of these, 802.16 has come to be known as *WiMAX*.

The WiMAX family of standards are part of the IEEE 802.16 standards shown in Box 'WiMAX specifications', and are mostly associated with IEEE 802.16e. Table 5.8 lists the major parameters of the system. Again, WiMAX is relatively unknown compared with WCDMA, and a list of prominent players in this field is shown in Box 'WiMAX players'. While IEEE 802.16e is not yet finalised, a number of vendors supply 'WiMAX-Ready' equipment.

WiMAX specifications

Three major IEEE 802.16 standards have been developed: WiMAX is generally synonymous with 802.16e (with mobility).

	802.16	802.16d	802.16e
Completed	Dec. 2001	Late 2004	Late 2005
Bit rate	32–134 Mbps at 28 MHz	Up to 75 Mbps at 20 MHz	Up to 15 Mbps at 5 MHz
Modulation	QPSK, 16-QAM, 64-QAM	OFDM 256 BPSK, 64-QAM	OFDM 256 BPSK, 64-QAM
Mobility	Fixed	Fixed and portable	Mobile (limited roaming)
Bandwidth (MHz)	20, 25, 28	1.25–20	1.25–20

WiMAX players

WiMAX technology is being standardised through IEEE 802.16. **WiMAX Forum is** an association of the WiMAX players: developers, manufacturers, operators and academia.

Intel: is one of the main supporters of the WiMAX standards, and has announced that it will include WiMAX transceivers in its chipsets.

Manufacturers: A number of manufacturers are active in the field, already producing pre-WiMAX equipment. One such manufacturer in **Alvarion**.

Operators: A number of operators are examining WiMAX. One such operator is Yozan in Japan.

Table 5.8 WiMAX parameters

Bandwidth (MHz)	1.25, 5, 10, and 20
Sub-carriers	16, 64, 128, and 256
Symbol length	102.4 μs
Guard-time	12.8 μs
Modulation	QPSK, 16-QAM, and 64-QAM
Channel coding	Turbo and LDPC

5.3.4.1 Peak throughput

Downlink peak throughput for the WiMAX system is calculated in a similar way to that of WCDMA HSDPA (see Box : Peak throughput). Again, peak throughput is calculated where interference power is minimal, that is, areas close to the transmitter. As WiMAX specifies several carrier bandwidths from 1.25 to 20 MHz, different throughput figures result. Figures for a system with 5-MHz bandwidth is provided. These figures can be scaled to find the figures for systems with different bandwidths.

Using the parameters of Table 5.8, the peak throughput of WiMAX system for both uplink and downlink are calculated in Box 'WiMAX peak throughput'.

WiMAX peak throughput

An example for a 5-MHZ WiMAX system is calculated below.

System bandwidth	5 MHz Mcps
Sub-carrier frequency spacing	9.766 kHz
Number of sub-carriers	384
OFDM symbol duration	115.2 μs
Symbol rate per sub-carrier (1/115.2 μs)	8680
Total symbol rate (8680 x 384)	3.3 Msps
Information symbol ratio (80%)	2.67 Msps
64-QAM modulation: 6 bits/symbol	16 Mbps
Channel coding rate 3/4	12 Mbps
Peak throughput	12 Mbps

WiMAX uses OFDM also in its uplink. This can lead to problems with power amplifier design with regard to the PAPR problem as discussed on page 38. Since commercial WiMAX systems are yet to be introduced to the market, it remains to be seen how much of the promised benefits can in fact be delivered in both downlink and uplink.

5.3.4.2 Average throughput

Typical average throughput figures for WiMAX are based again on the different possible levels of SINR present within a cell area. This author's optimistic estimation of average throughput for a pre-WiMAX 802.16d system with a frequency reuse factor of 1 is shown along side the throughput figures, as well as frequency-utilisation efficiency figures, for downlink in Table 5.9.

5.3.5 DoCoMo test system

The leading Japanese operator NTT DoCoMo has been one of the first organisations to start research on 4G systems. They have built a prototype of an OFDM-based broadband wireless communications system capable of delivering up to 300 Mbps in the downlink and 40 Mbps in the uplink. Moreover, the system can deliver up to one giga bit per second using a 4×4 MIMO antenna system. The system parameters are shown in Table 5.10, and throughput figure in Table 5.11. The system uses a wide bandwidth, 100 MHz, in the downlink and 40 MHz in the uplink, and can deliver up to four parallel high definition TV quality streams in the downlink. The average throughput efficiency is expected to be similar to 802.16 WiMAX: 0.6 bps/Hz (from Table 5.9).

WCDMA HSDPA with OFDM and 64-QAM

DoCoMo, among a number of other organisations, is proposing to include an OFCDM technique within the HSDPA specifications. This is still a study item. Furthermore, a 64-QAM modulation mode is also being studied. These are proposed in order that the air-interface specification of WiMAX and WCDMA can be harmonised.

Table 5.9 Throughput figures for WiMAX downlink

	WiMAX downlink throughput	
	(Mbps)	(bit/s/Hz)
Peak	12	2.4
Average	3	0.6

Table 5.10 DoCoMo broadband wireless system parameters

Bandwidth	100 MHz
Sub-carrier spacing	131.836 kHz
Sub-carriers	768
Symbol length	7.585 μs
Guard-time	1.674 μs
Modulation	QPSK, 16-QAM, and 64-QAM
Channel coding	Turbo, rate = 1/8–8/9
Pilot allocation	10%

Table 5.11 Downlink throughput figures for DoCoMo system

	WCDMA HSDPA (Mbps)	OFDM downlink throughput (bit/s/Hz)
Peak	310	3.1
Average	60	0.6

Table 5.12 Transmission power limitations for the WLAN systems' unlicensed frequency bands in several countries/regions

Frequency band (MHz)	US (dBm)	Europe (dBm)	Japan (dBm)	China (dBm)
2400–2483	30	20	24	20
5150–5250	23	23	22	–
5250–5350	30	23	–	–
5470–5745	–	23	–	–
5725–5825	36	–	–	27
5825–5925	–	–	–	27

5.3.6 Wireless LAN throughput

As described in Section 1.8.4, WLAN systems use a CSMA mode for accessing a common frequency resource. Frequency bands used for WLAN are of the unlicensed type, and are listed in Table 5.12. This means that the transmission power for these systems are limited by government regulations to low levels (these are listed in the table). Low levels of transmission powers result in small coverage areas, as discussed in Section 2.9. As WLANs were initially designed for small-area coverage, this was not considered an issue. However, as WLAN devices position themselves to compete with broadband wireless systems, they need to use licensed bands in order to increase their transmission power and broaden their area of coverage.

5.3.6.1 IEEE 802.11a

There are several WLAN standards. Here, we will use the IEEE 802.11a standard to illustrate throughput figure calculations for WLANs. The IEEE 802.11a standard uses an OFDM modulation, consisting of 52 carriers, each with a bandwidth of 312.5 kHz, within a bandwidth of 20 MHz. The standard specifies adaptive coding from rate 1/2 to rate 3/4, and adaptive modulation from BPSK to 64-QAM. Transmission rates are adopted from a White paper by Atheros Communications, and are summarised in Table 5.13. The average value is calculated over a radius of 68 m. Note that, as this is a CSMA system, the uplink and downlink are hard to distinguish. The throughput figures are total sums for two-way communications.

Table 5.13 IEEE 802.11a throughput figures

	Throughput (Mbps)	Efficiency (bps/Hz)
Peak	27.0	1.35
Average	9.4	0.47

Owing to the CSMA nature of the IEEE 802.11 technology, the maximum throughput figures can dramatically decrease as the number of users connecting to an access point increase. This is mostly due to inefficiencies associated with multiple users trying to access the same resource and backing off as they sense the channel is busy. Collisions may also occur, as a device may not correctly detect that the channel is actually busy. A number of collision avoidance (CA) algorithms exist to reduce this probability.

5.4 Spectrum Cost

Large amounts of money were paid for 3G spectrum licenses in many countries across Europe and North America. These licenses raised hundreds of billion dollars for governments. However, the excessive sums paid led many an operator to the brink of bankruptcy. This is cited as one of the reasons why the Internet and Telecommunications industry (IT) bubble burst in the year 2000. Meanwhile a number of countries, including Japan, opted for a different approach, the so-called 'Beauty Contest', to allocate spectrum to prospective operators. While those whose bids 'won' in the auction process have struggled to establish 3G networks, the operators who received licenses through beauty contests have been relatively much more successful. Likewise, where governments did not raise money from the auction process, they seem to have gained more from the successful operators, who paid more taxes.

It remains to be seen how licenses will be allocated for wireless broadband services. To date few examples exist. One example, however, is the Republic of Korea, where licenses in the 2.3-GHz band have been granted through a beauty contest at a nominal fee. This fee is calculated on the basis of the number of subscribers to these services. In the subsequent text, we will discuss briefly the cost per subscriber for 3G licenses and what these may become for broadband wireless services.

5.4.1 Cost per subscriber

The costs of spectrum per subscriber are shown in Table 5.14 for two countries: the United Kingdom, where the auctions were held for the 3G spectrum, and in Japan, where licenses were granted without charge to existing operators and they are subsequently charged a nominal fee per subscriber for their spectrum utilisation. These figures are for 3G spectrum. The table shows the number of subscribers all the operators combined can support using services under their licenses. In the United Kingdom, these figures yield the cost per subscriber: a sum that should be recovered over the period of the validity of the license lease, assuming a 10-year payback period, with zero interest rate. In Japan, where spectrum

Table 5.14 License fees, subscriber numbers, and cost per subscriber per year (paid over a 10-year period)

Country	Money paid	Estimated subscribers	Cost per subscriber
UK	$29 B	15 M	$193
Japan	–	32 M	$36

was allocated to operators without an auction, the cost of spectrum is the fee that operators must pay to the government.

It is interesting to note that where the cost of spectrum is fixed, as in the United Kingdom, there is great pressure on the operators to increase their subscriber base. This will drive the cost per subscriber down, and allow the operator to provide services at increasingly reduced cost to the end-user. Such an operator must use the most frequency efficient technology to try to serve the largest possible number of subscribers.

In contrast, when the cost of spectrum is variable, as in Japan, the operator is not under great pressure to recoup the initial investment in spectrum acquisition. However, as network equipment costs are fixed, this means that the operator must try to serve end-users efficiently. However, if the government can supply the operator with extra spectrum as initial allocation is exhausted, and if this does not require a major upgrade of network equipment, then frequency-utilisation efficiency is not a major concern.

As mentioned above, these remarks are for 3G spectrum. It is likely that in many countries, the spectrum allocation processes for broadband wireless systems will not be based on auction, and rather follow a path similar to Japan. Even auction prices will probably not go so high as in the case of 3G systems. In any case, it does appear that the cost of acquiring a spectrum license will not be the major cost in operating a network.

Summary

In this chapter, we have discussed the issue of spectrum cost for providing a certain grade of service. We explained the efficiency of different access technologies in utilising frequency. Then, we used frequency-utilisation efficiency to calculate the relative cost of spectrum for different technologies in providing broadband wireless services. Should the cost of spectrum be fixed, as in spectrum acquired through auction process, then frequency-utilisation efficiency is of great importance. If the cost of spectrum is variable, that is, a fixed amount per subscriber, and if a government supplies the operator with 'unlimited' spectrum, then the efficiency of technology is not so important.

Further Reading

- On WCDMA:
  - 3GPP web site: `http://www.3gpp.org/`.
  - UMTS Forum web site: `http://www.umts-forum.org/`, 2005.

- On WiMAX:
  - 3GPP web site: `http://www.3gpp.org/`.
  - WiMAX Forum web site: `http://www.wimaxforum.org/`, 2005.
- On TD-CDMA:
  - 3GPP web site: `http://www.3gpp.org/`.
  - The Global UMTS TDD Alliance web site: `http://www.umtstdd.org/`, 2005.
- On Wireless LAN:
  - White Paper, *Measured Performance of 5-GHz 802.11a Wireless LAN Systems*, Atheros Communications Incorporated, 2001.
  - 3GPP web site: `http://www.3gpp.org/`.
- Voice codecs AMR web site and ITU standards
- Traffic models ITU:
  - One typical model is described in Selection Procedures for the Choice of Radio Transmission Technologies of the UMTS, UMTS Technical report TR 101 112.

6

Cost of Equipment

As discussed, wireless telecommunications is by and large a fixed cost business. In Chapter 5, we discussed the major initial element of fixed cost: that of frequency spectrum and its acquisition. Another major cost is the wireless network infrastructure. Figure 6.1 is a diagram of various typical equipment in a typical GSM wireless communications system. The infrastructure cost is associated with purchasing and installing these equipments; in other words, it is the total cost of the equipment used to transmit data between an end-user mobile device and the public network domain.

The greater part of this cost is associated with base station equipment. Base stations connect the wireless world to the fixed, wired world, converting radio signals to digital bit streams that can be transmitted over fixed links. As a very large number of these base stations are required to provide adequate coverage over a region's populated areas, it logically follows that base stations are by far the largest part of the infrastructure cost. Next in the cost ranking are network nodes that control these base stations. They are usually called *radio network controllers (RNCs)*. The structure of these devices and their interface with the base stations are well defined. The cost per RNC may be higher compared with that of a base station, but as long as fewer units of RNCs are required (each RNC can control from a few to tens of base stations), the total RNC cost is less than total base station cost. There are several more nodes that interface legacy systems: these are known as gateways. An example is the gateway to the public analogue telephone network. There further exist several other nodes associated with authentication and billing. All of these nodes are connected with each other using high-speed, fixed, digital links such as optical fibres. The only exceptions are (mainly) remote base stations, which may be connected to an RNC using point-to-point wireless links.

On the other side of the wireless link is the end-user equipment. The best example is the mobile phone, the ubiquitous device of the first-, second-, and third-generation mobile communication devices. In this book, we have referred to these as mobile stations. One expects that end-users should bear the cost of these devices. Ultimately, end-users must pay not only for their own personal mobile phones but also for all the other costs of the

Broadband Wireless Communications Business: An Introduction to the Costs and Benefits of New Technologies Riaz Esmailzadeh

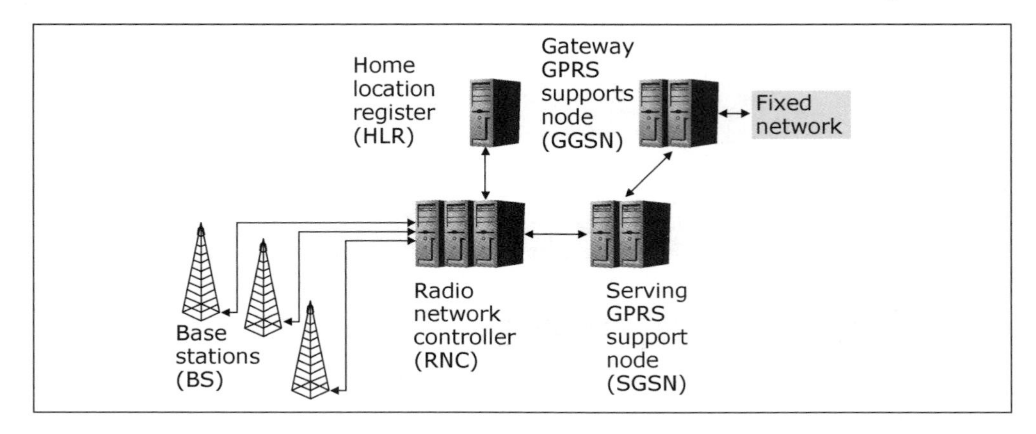

Figure 6.1 Infrastructure nodes

operation if the operation is to be viable. The existence of subsidies in most parts of the world seems to suggest that only the operator must bear the large initial cost.

This chapter discusses costs associated with the infrastructural equipment of a wireless communications network, as well as the cost of procured mobile stations. We provide a survey of the different nodes (a node referring collectively to any equipment in the chain, starting with a mobile station to gateways to the internet, and on to public networks), a summary description of a node's different elements and their relative costs, and how these costs have been changing as a result of technological development. Throughout, we try to give, where possible, actual cost figures. We note however that in many instances actual cost figures are closely held vendor/operator secrets. These figures are usually negotiated quite rigorously between an equipment vendor and an operator; the final figures are a function

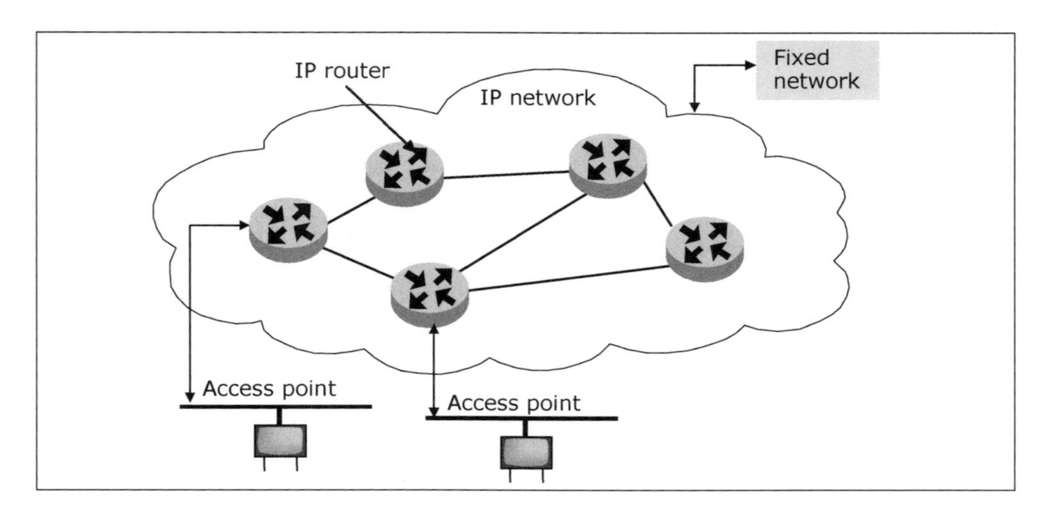

Figure 6.2 A Wireless LAN network

of the transaction volume as well as the balance of power between the operator and the vendor. The actual figures may have dubious values themselves as the prices are constantly changing. One exception, of course, is the cost of wireless LAN equipment, which are retail commodities, and prices are in the public domain. We start with a discussion on the costs of the base stations, and continue with other network equipment and mobile stations; the cost of Wireless LAN nodes will be addressed as well (illustrated in Figure 6.2).

6.1 Base Station Structure

A base station is the device connecting the wireless domain to the fixed, wired domain. On the one side, it performs the functions associated with transmitting and receiving radio signals – coding, modulation, amplification, and transmission. One the other side, it connects to a fixed link, which transfers data to and from a public network. A large number of base stations are needed to provide coverage to the populated areas in a region. For example, it is estimated that some 20 000 base stations are deployed in Japan to provide coverage for the areas that together host over 99% of the Japanese population. Several different types of base stations exist and, as such, each carry a different price tag. Still, their internal structure is the same. Figure 6.3 illustrates different parts of a base station as specified by the Open Base Station Architecture Initiative (OBSAI).

In general, a base station comprises a radio frequency (RF) module, which interfaces the wireless domain; a baseband (BB) module, which processes information signals; a transport module, which interfaces the fixed network; the control module, which delivers necessary signals, such as timing information; the system software, which controls the total system and a power supply. Technological developments have aimed at enhancing the performance of all these components. However, the degree of enhancement has been different owing to each component's physical characteristics; and future enhancement is also expected to be of different degrees. The extent to which each new wireless technology utilises, and affects, each of these integral parts is also different. We define in the following text, the function of each component, its recent technological improvements, and what significant role each

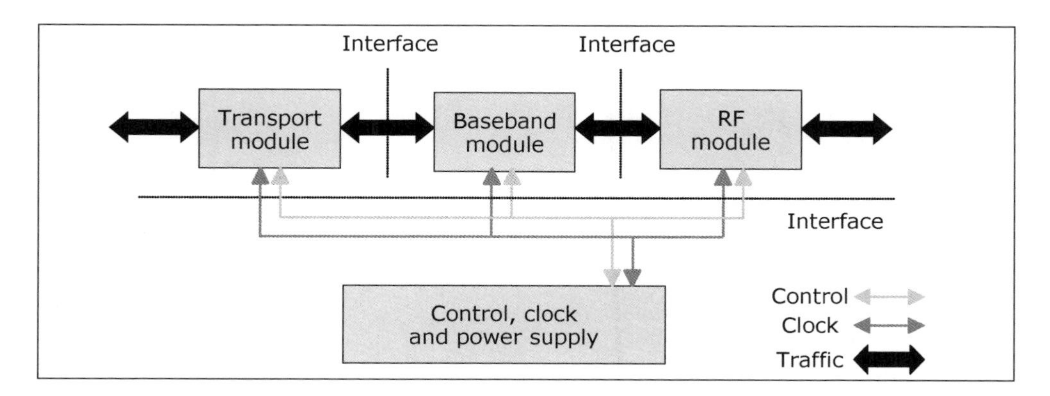

Figure 6.3 A base station modular structure as defined by OBSAI. Reproduced by permission of OBSAI

one plays in a wireless system. These components are typical of public wireless standards such as WCDMA, TD-CDMA, CDMA2000, and WiMAX.

6.1.1 RF module

The RF module is the interface to the wireless domain, as illustrated in Figure 6.4, and includes the following:

- Antennas: These may be omni-directional, sector, adaptive array, or MIMO antennas.
- Antenna control units: The control units are used particularly for multiple element antennas such as those with adaptive arrays and MIMO.
- Mixers (up/down convertors): These devices convert the transmission and reception signals from RF frequencies (frequencies of interest to this book are in the 2 GHz–5 GHz range) to baseband (several MHz).
- Filters: These are used to filter out unwanted signals during both the transmission and reception processes.
- Power amplifiers: These devices increase the power level of transmission signals to ensure the delivery of desired signal quality to the mobile stations.

Antennas

As discussed in section 3.4, advancements have been made in the field of antenna technology. As might be expected, improved antennas cost more. Figure 6.5 shows two typical antennas used for present mobile communication standards: a three-sector antenna, and an eight-element adaptive array antenna.

Antenna costs are in the order of a few hundreds of dollars. As the antenna material cost is minor compared with the antenna's processing RF modules, the cost of adaptive array antennas can be significantly higher. These depend on the type of processing: whether permanent beams or stirred beams are used. The former requires processing functions at

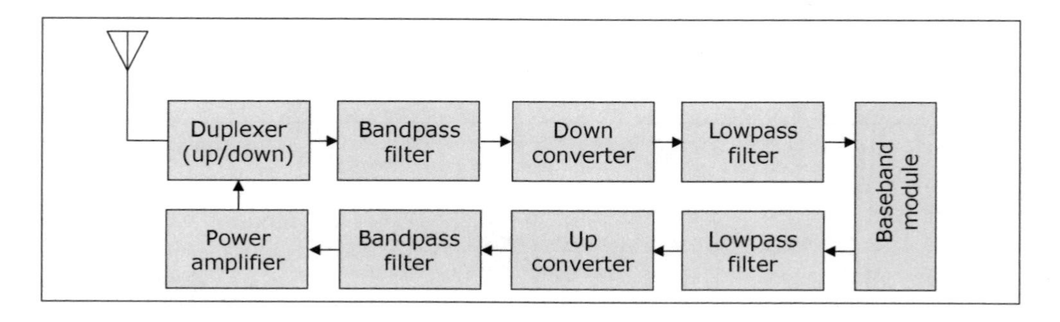

Figure 6.4 RF module

Figure 6.5 Three-sector and eight-element adaptive array antennas for base stations

the baseband, and therefore is relatively inexpensive. The latter requires processing in RF, and the cost of the antenna including the control module can be 10–100 times higher. While figures for sector antennas are openly available, adaptive array antennas' cost figures are not.

Some broadband wireless technologies have integrated the adaptive array antenna function into the base station receiver. These base stations are more complex, but the resulting combined product can be more economical.

As antennas are external to base stations, an operator may purchase both omni-directional and sector antenna modules independent of a base station vendor. These antennas are integrated with the base station at the time of installation. An exception is with the case of adaptive array antennas, which require special RF processing functions. These are usually developed by the vendor and are included in the base station equipment. The antennas themselves can be provided by the vendor or procured independently according to defined specifications. Systems using adaptive array antennas or MIMO need relevant control algorithms integrated into the control software, and their implementation needs to be well coordinated with the base station equipment vendor. The higher costs are well compensated by the performance enhancements as discussed in section 3.4.

Mixers

These devices transfer a signal from its baseband (unmodulated) state – centred at 0 Hz frequency with a bandwidth of usually several MHz – to a signal centred at the carrier frequency, and vice versa. The transfer in the earlier 1G and 2G base stations was done in several stages. Several intermediate frequency conversions, with associated filtering and amplifications, were carried out before the signal arrived at RF band. Advances in electronics and updated base station design have made it possible now for this process to be carried out with a single conversion. The new process is known as 'direct conversion'. The cost of these newer devices is small – and would be the same for both CDMA-based and OFDM-based standards as the necessary processes are the same.

Filters

Filters are used to block unwanted signals from transmission and reception. As shown in Figure 5.1, the frequency spectrum is shared among many services. Interference from devices operating in adjacent frequency bands can be high if proper filtering is not provided. Therefore, all standards define and ensure that spurious signal levels, outside their operating bands are below a certain level.

A typical filter mask is shown in Figure 6.6. Such a filter ensures that spurious interference outside the operating band remains below a designed level. Even so, such a filter does not by itself ensure that interference to neighbouring services is below a satisfactory level. Further filtering may be required to ensure all present services are unharmed by the introduction of a new service. This usually occurs when two systems need to operate in close proximity to each other, using adjacent or nearby bands.

An example of this is the coexistence of WCDMA and TD-CDMA services in Japan. Here, WCDMA services are offered in the 1920–1980 MHz (uplink) and 2110–2170 MHz (downlink) bands. TD-CDMA services are to be offered in the 2010–2025 MHz band. Table 6.1 lists the interference levels caused by the two systems. Although filter specifications for both standards ensure a certain degree of mutual isolation, extra filtering would be required if the base station antennas are co-located, (for example, on the same roof (~5 m apart), or are close to each other (for example, on top of neighbouring buildings). Extra filtering ensures that the two systems can operate within the same geographical area. These filters are usually external to the base station, and are physically placed just below the antenna.

The cost of typical filters again is a function of volume, and is quite small. Furthermore, the cost is expected to be similar for both CDMA-based and OFDM-based standards, because of the similar functionality.

Power amplifiers

Power amplifiers are needed to increase the milliwatt-level output signal from mixers and filters, and amplify it to several watts. This ensures that the received signals at all

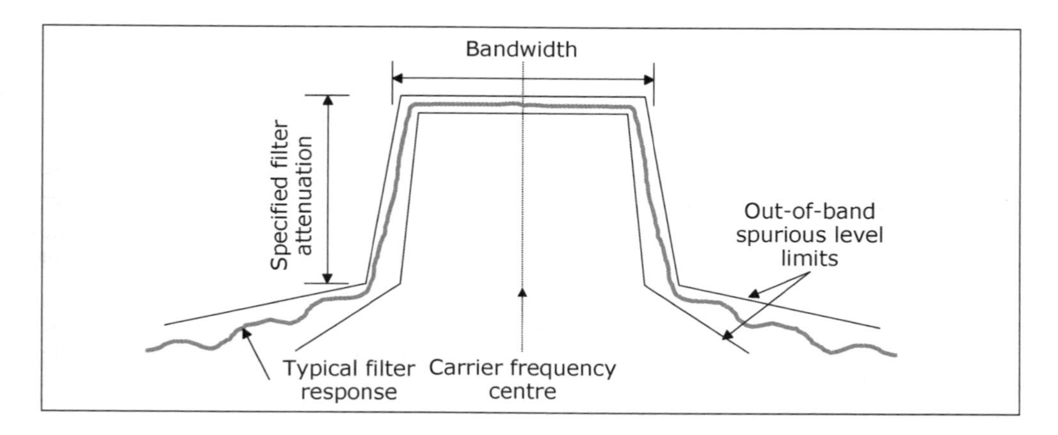

Figure 6.6 A typical filter mask

Table 6.1 Interference from a TD-CDMA base station (transmitting in the downlink) to a WCDMA base station (receiving in the uplink)

WCDMA BS receiver frequency band	2110–2115
TD-CDMA BS transmitter frequency band	2010–2025
TX and RX bandwidth	3.84 MHz
Interference from TD-CDMA band to WCDMA band	−43 dBm/3.84 MHz
Antenna to antenna path loss (3 m separation)	50 dB
In-band interference level	−93 dBm
WCDMA receiver acceptable power level	113 dBm/3.84 MHz
Extra filter required	20 dB

Table 6.2 Relative power amplifier costs

Power amplifier output level	Typical cost $
100 mW	<10
1 W	100–300
10 W	1000–2000

places within the coverage area have a sufficient SNR level. A good power amplifier can be defined as having a constant gain within its operating range. Further, it is desirable that the device has a wide operating range, that is, it should provide the same constant gain over a large range of input power levels. Furthermore, such devices are required to have 'low noise' figures to add minimal additional noise to the system. High linearity, a wide range, and low noise are the highly desirable characteristics of the power amplifier.

High-quality performance requirements for these devices make them quite voluminous and expensive. A base station power amplifier may have a size of up to ten litres. Cost increases with the device's maximum output power and linearity. Typical cost figures for some output amplifiers are shown in Table 6.2. The price is very much a function of the power amplifier's output power and linearity characteristics. The figures presented here are this author's estimates.

Yet another important aspect of power amplifiers is their operating efficiency. This is defined as the amount of power used for amplification compared with the total power consumed by the device. The difference is converted to heat, which must be dissipated through heat sinks.

The operating range of a power amplifier is illustrated in Figure 6.7. An ideal linear power amplifier delivers a constant gain up to the point of saturation. If the input power level is constant at all times, then the amplifier can operate at 'knee point', that is, its maximum output power. This will result in maximum operating efficiency, and the least generated heat. Operating at maximum efficiency also causes minimum out-of-band interference. If the input signal is any larger than the maximum input power level, then the signal will be clipped, resulting in nonlinearity and signal distortion, and leading to increased reception errors.

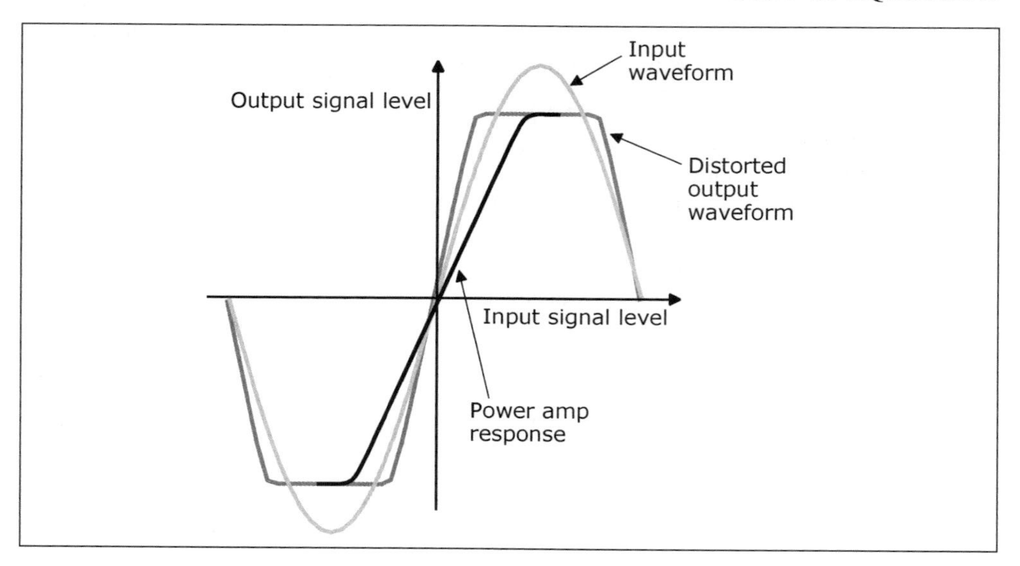

Figure 6.7 Operating range of a power amplifier and signal clipping

Table 6.3 Input-back-off and efficiency for a Class-F power amplifier used for CDMA- and OFDM-based technologies

Standard	Required back-off (dB)	PA efficiency (%)
No input back-off	0	60
CDMA	3	44
OFDM	6–9	30

The input signals delivered to a power amplifier naturally do have a certain degree of power level variation. For example, multi-level modulations used in 16-QAM systems or multi-carrier modulation used in OFDM systems, lead to power variations of several dB. (See PAPR discussion in Section 2.14). In order to avoid clipping and signal distortion, the systems cannot operate at maximum input power levels. Thus, the input is backed off by several dB from the maximum possible input level to ensure that linearity is maintained. This practice is known as input-back-off (IBO).

One problem with IBO is that the efficiency of the power amplifier is reduced. This results in greater heat generation, larger power consumption, the need for larger heat sinks and more complex base station mechanical design. In Table 6.3, some wireless communication technologies are compared with respect to their required IBO for a Class-F power amplifier. (See 'further reading' list at the end of this chapter.)

The cost of power amplifiers has been decreasing while more advanced devices have been produced and market size has increased. The cost decreases, however, have been incremental, and it is expected that the cost of linear power amplifiers will remain significantly high in the foreseeable future.

Total RF cost

The cost of an RF module can be expected to be similar for all broadband wireless systems discussed here, given that the same power amplifier classes are used. Different power amplifier back-off levels result in different heat dissipation requirements, and base station size. If similar output power levels are required, then different amplifiers are needed, and different system costs result. Moreover, any improvement in the budget link due to processing, for example, joint detection, or better antennas, for example, adaptive arrays, can reduce the burden on the power amplifier, and reduce the cost of the RF module. The TD-CDMA system's use of joint detection can therefore yield reductions in base station cost. Generally, the lower the transmitter power is, the lower the device cost can become. TDD-based systems also benefit from the fact that duplex switching in the time domain is simpler, and therefore cheaper than in the frequency domain. Furthermore, a single set of filters and oscillators are required as uplink and downlink communications are carried out in the same frequency band.

6.1.2 Baseband module

Within this module, conversion is made between the digital bit stream of the fixed network domain, and the analogue radio signals of the transmitted signals. On the transmitter side, the incoming bit streams are encoded, modulated (converted from digital to analogue), and passed onwards to the frequency up-convertors of the RF module. On the receiver side, the baseband converted signals from the RF module are demodulated and decoded (and in the process converted to digital) to produce a bit stream that can be sent to the fixed network. These processes are illustrated in Figure 6.8.

Baseband procedures involve considerable signal processing. Relatively speaking, however, for all wireless communication technologies, the signal processing requirements for the transmitter side are not large. Both encoding and modulation processes involve only a few multiplications and additions and therefore can be carried out with minimum microprocessing. On the receiver side, however, there is a need to carry out all of the equalisation processes discussed in Section 2.3. These involve a large number of complex multiplications in real time, and result in extensive signal processing.

There is a significant difference between the signal processing requirements of different broadband wireless technologies discussed in this book. In particular, TD-CDMA with joint detection is highly processing-intensive. A comparison on the signal processing requirement

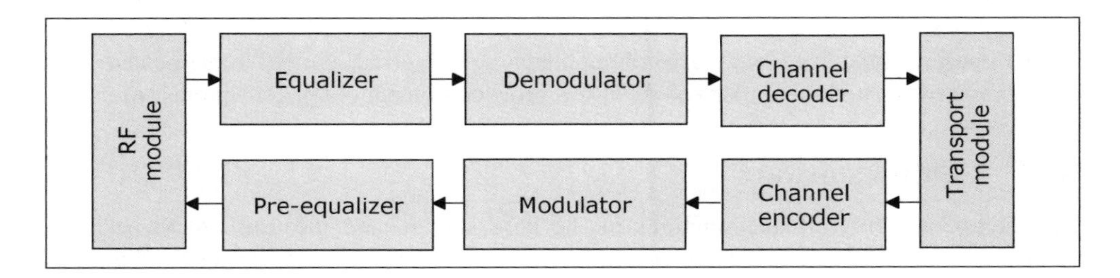

Figure 6.8 Baseband processes

Table 6.4 Processing requirements for WCDMA and OFDM receiver structures, relative to 'one addition' process. Reproduced by permission of 3GPP

Technology	Relative process count
RAKE for WCDMA release 99	~800
WCDMA OFDM HSDPA	~1300
RAKE for HSDPA (15 codes)	~1800
Joint detector for HSDPA (length 16 chips)	~7800
Joint detector for HSDPA (length 32 chips)	~24 000

of different devices is shown in Table 6.4. This table shows that an OFDM-based receiver can be produced with relatively smaller processor power, leading to lower costs in the development of this baseband unit.

One can conclude from Table 6.4 that different technologies can have a different degree of associated costs as a result of their different degree of processing requirements. Ultimately though, the costs are likely to be very similar. As all of these technologies will carry out baseband processing on dedicated application-specific integrated circuits (ASICs), the cost of manufacturing will be similar. Another important factor is that processing power has been following Moore's law: that is, the doubling of processing power every 18 months. This means that increasingly complex baseband processing will become possible at lower and lower costs. An example that demonstrates this is the TD-CDMA devices, which already have complex joint detection functions implemented in ASIC.

What distinguishes the technologies more is the cost of licensing of different technologies used within the receivers. This issue will be addressed in a later section of this chapter.

6.1.3 Transport module

This module interfaces the fixed-line network. The type of interface depends upon the transport protocol. In earlier 1G and 2G systems, the backbone transport used asynchronous transfer mode (ATM) links. Initial 3G systems, such as WCDMA also used ATM. However, the newer releases of 3G standards, the super-3G systems and the proprietary broadband wireless systems are designed to be all-IP. The transport module, therefore, arranges the bit stream from the baseband module into internet traffic and transports the packets to the ethernet links.

All transport modules largely use common techniques and are simply implemented. The cost is rather low, and is similar for all of the broadband technologies of interest to us.

6.1.4 Control software

The module controls all the functions of the base station, and the traffic flow between modules. The 3GPP standard keeps most of the network control functions within the radio network controller, and the base station is by and large a transceiver. However, control of base station functions, such as traffic flow between modules and power management, are necessary. These are carried out by this module. The modular base station design enables larger operators to procure modules from a number of vendors, and integrate the system themselves, using their own control software. This can result in significant cost savings. Smaller operators usually purchase base stations, including the control software, from manufacturers. Such operators cede significant system and software upgrade control to the manufacturer.

The network architecture of IEEE standards systems (e.g. 802.11) are not modularly defined. However, in time a modular approach is expected, and a similar situation to 3GPP systems emerge. But meanwhile, the manufacturer delivers a full integrated system to the operator.

Overall, the modular approach is beneficial to operators, who gain some measure of bargaining power through multi-sourcing.

6.1.5 Clock and power supply module

The clock module ensures that all receiver and transmitter units within the RF module are synchronised with each other and with the baseband module. There have been significant improvements in the accuracy of systems clocks, allowing for nano-second precision of synchronisation clocks. This is an important function, increasingly so for broadband wireless systems with very high transmission rates.

Some standards, such as IS-95 and TD-CDMA, require the whole system to be fully synchronised. Synchronisation is acquired and maintained through geostationary-positioned satellite (GPS) systems. The clock module for these systems is necessarily more complex and therefore more expensive.

Power supply systems convert the public AC supply to a DC supply suitable for the RF module. The costs for this module is expected to remain stable, be a function of volume, and be similar for all of the broadband systems of interest here.

6.1.6 Device volume

Base station volume has been getting smaller with each new generation and release. In particular, the size of baseband units has been decreasing steadily as more and more processing power is delivered by smaller ASICs. Table 6.5 shows how the typical size of a WCDMA base station has changed since its initial release in 1999, with pictures of the two base stations shown in Figure 6.9. The smaller volume brings significant flexibility to network design. Older, larger base stations usually need special rooms at the tower site to accommodate them. These might be situated some distance from the antenna, thus leading to substantial cable requirements. Smaller base stations can be placed on rooftops, or even attached to electricity poles, reducing installation cost. Furthermore, air-conditioning cost to remove the heat generated by base stations has been quite significant. Smaller designs do mitigate these costs. Nevertheless, some physical characteristics are still the same, such as the size and types of antennas.

6.1.7 Repeaters

A repeater's function consists mainly of receiving a signal at one of its antennas, amplifying it, and transmitting it out from another of its antennas. As such, it has a very simple structure;

Table 6.5 Changes in the size and relative cost of a WCDMA base station

Model (year)	Size (in litres)	Relative cost	
BS2001	900	1.0	
BS2006	27	0.1	Outdoor macro-BS
BS2007	54	0.1	Indoor micro-BS

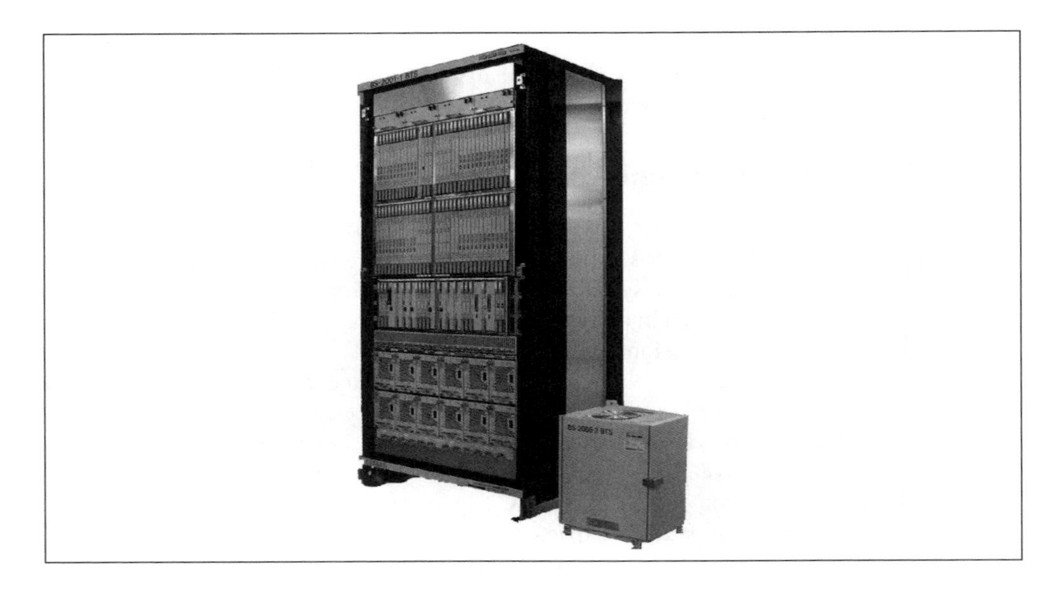

Figure 6.9 Two WCDMA base stations, BS2001 and BS2006. Reproduced by permission of ITMedia Mobile

a repeater does not carry out baseband processing, and therefore has no baseband module. It does not require extensive control software. Antennas and RF modules are also smaller and less expensive. The costs for a repeater are expected to be similar for all of the broadband wireless technologies of interest here.

6.1.8 Base station supply chain

We have reviewed the list of components, hardware and software, that comprise a base station. A base station vendor is usually the final player who assembles components from a large group of suppliers. A typical supply chain is shown in Figure 6.10. The vendor does design and develop some of the components, mainly the baseband ASICs and RF components. And, the vendor certainly develops the control software for smaller operators. But the filters, power amplifiers, clocks, power supplies, Ethernet interfaces, and so on, are usually procured from third party suppliers. The vendor may or may not themselves supply the antennas as discussed above. Any control software and equipment for antennas that are required are delivered in a base station package. In many cases, the vendor is

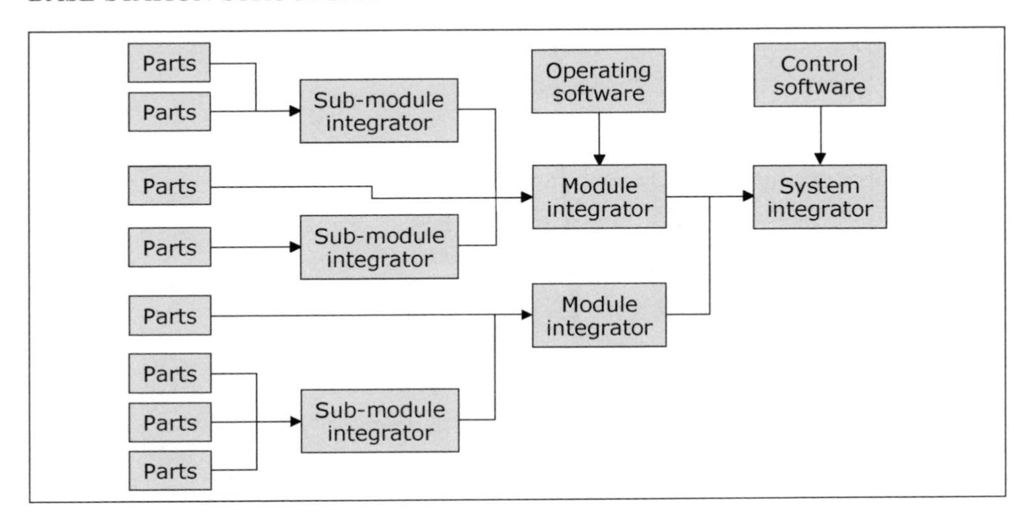

Figure 6.10 An illustration of infrastructure development supply chain

also the system integrator; the vendor supplies the base station, installs and tunes the antennas, installs control software, and ensures that all modules work correctly. Such a vendor obviously holds significant bargaining power over the operator.

The modular design of the base station enables an operator to use different suppliers for different modules. Independent system integrators, who can be the operator's subsidiary, integrate and fine-tune the network. In such a case, the operating agency itself develops and updates the control software. Japanese NTT DoCoMo is an example of such an operator. It specifies in detail each module and then has it manufactured by multiple vendors. Clearly, DoCoMo exercises great power over its vendors. Nevertheless, DoCoMo is an important customer: Figure 6.11 shows DoCoMo's capital expenditure over the recent five years. A total of \$20 billion has been spent in 3G infrastructure build-up to provide a coverage of 99.7% of the Japanese population. Generally, the highest quality of supplied modules are expected, and the manufacturer who does not deliver good reliable products is accordingly left out of the list of future suppliers. This has conferred DoCoMo with a great reputation for reliability in service in Japan.

6.1.9 Licensing cost

What perhaps distinguishes the cost of different broadband wireless technologies from each other most is their licensing costs. CDMA-based standards are based on many patents owned by Qualcomm Inc., who developed the CDMA technology for IS-95 standard. Many of these technologies are also necessary in WCDMA and TD-CDMA technologies. These standards also need many other new technologies, developed since IS-95. A large group of companies (and other institutions such as universities) own intellectual property rights (IPR) in the 3G CDMA systems. To ensure fair licensing, many of these companies jointly license their patent through a 3GPP body. Other companies have also promised to work to keep the total royalties to below a fair level. It is estimated that 10% of the sale price

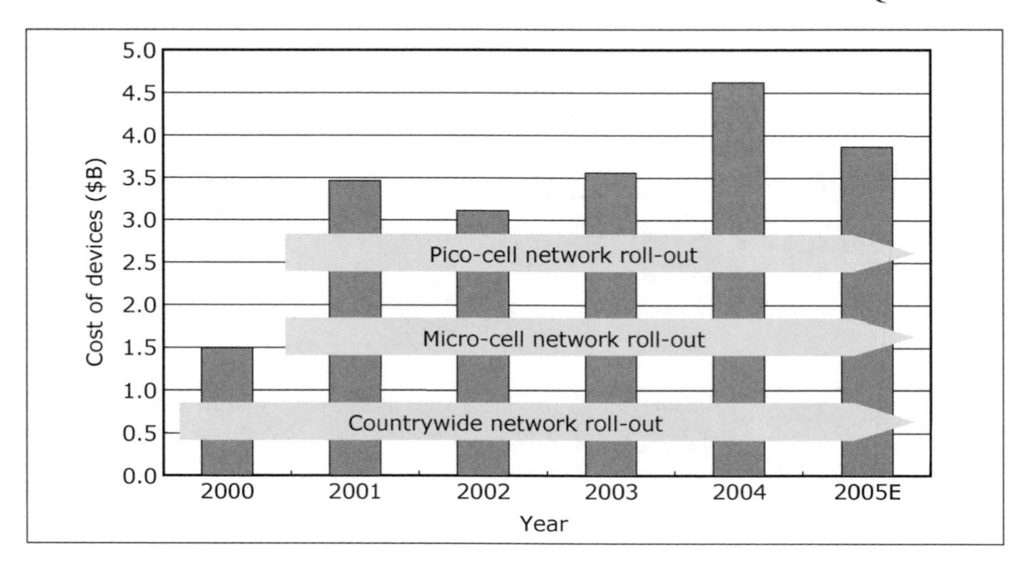

Figure 6.11 DoCoMo mobile network capital expenditures

for WCDMA equipment (both base stations and end-user equipment) is paid to patent holders. It should be noted that a large part of these patents are for call establishment, and network control purposes. Pure physical function IPRs, such as coding and modulation, are in contrast very few.

It should be noted that, as compared to WCDMA, TD-CDMA is less dependant on CDMA patents, and therefore IPR cost should be relatively lower.

The IEEE standards boast of a relatively lower IPR barrier. OFDM technologies are well known, and are transparent. Moreover, because IEEE standards only define layer 1 and 2 issues, much of the network and traffic controls are unspecified and left to the vendor and operator to handle. All these result in a lesser number of patents that need to be licensed. However, should WiMAX and WLAN networks be designed to provide higher level functions such as full coverage and handover, then the relevant IPR (similar to those used in 3GPP standards) needs to be licensed. This means that licensing costs will become comparable with WCDMA.

Another issue in deciding on the licensing cost is what the end-user equipment is. Nowadays, many devices have wireless communications capability. A laptop PC and a mobile phone may have similar wireless connectivity characteristics. However, the 10% IPR rule does not apply equally to the retail price of these two devices. This is because a mobile phone is mainly a communications device, whereas a PC is used for communications as well as for many other applications. Royalty calculation is made based upon the percentage an equipment is used for communications.

6.2 IEEE 802.11 Wireless LAN Access Point

The price of IEEE 802.11 wireless LAN equipment, both the access points with IP router functionality and the end-user device cards, have decreased together with their shift in character from speciality devices to commodity products. Inclusion of the end-users' WLAN cards in many laptop computers as well as in the later embedment in Intel's Centrino processor have only helped them to spread more widely. The structure of these devices, however, are similar to those of the more complex standards such as WCDMA.

A wireless LAN access point (AP), shown in Figure 6.12 is a commodity product and its cost has been declining to less than $100. The size of the device has also diminished. The modules that go into a WLAN access point are the same as for a WCDMA base station, but on a much smaller scale. Because of the single user connection, and layer 1 and 2 functionality, the control software is much simpler. The RF module and baseband modules, similarly, are less complex. The size of an access point can be as small as one litre: similar to a pico base station.

Factors that make the cost of an access point so low are listed as follows:

- Small antennas
- Low-cost power amplifiers
- Simple power supply
- Single connection; no multiple access
- IP connection
- Layer 1 and 2 operation: simple control software
- Low licensing cost.

Figure 6.12 An IEEE 802.11 router-equipped access point

An IEEE 802.11 wireless LAN is designed to provide small coverage, and therefore, the rating of its components, as well as its operating software, are simple and low in cost. If IEEE 802.11 access points were configured to operate in a way similar to that of a WCDMA base station, with similar coverage and quality of service, then, similar antennas, RF modules, control software, and proprietary media access protocol algorithms would have to be provided. The addition of all these elements would bring the costs on par with the WCDMA or other public mobile communication systems. Note that licensing costs for many public systems are associated with higher layer functionalities, for example, those associated with medium access control. Because these functionalities are necessary for systems operating over a wide area, any enhanced wide-area 802.11 network can be expected to include them. This will result in increased cost for such devices.

The strength of the 802.11 standard is its independence from the high cost of modules for layer 3 and above. This is because 802.11 is designed to operate with 'best-effort' quality of service for small-area networking. Such a small-area business domain cannot be served quite as well using public technologies such as WCDMA. Thus, the end-user needs both of these kinds of services in combination. One popular solution being offered is a combination of these technologies in the form of dual-mode devices. We talked about this kind of service, a WCDMA-WLAN service offered by DoCoMo, in Box 'Mzone service' on page 87.

6.3 Network Nodes Costs

Network nodes relay information from a base station to a public network domain. Some major nodes for 3GPP are: (1) Radio network controller (RNC), (2) Serving GPRS support node (SGSN), (3) Gateway GPRS support node (GGSN) and (4) Home location register (HLR). A brief definition of these nodes is as follows:

6.3.1 Radio network controller

RNCs are devices that control the operation of several base stations and connect them to the public network. Their functionalities are beyond the scope of this book. Their structure and procedures are well defined and standardised as an integral part of WCDMA, TD-CDMA and CDMA2000 systems.

RNCs represent the most valuable part of a base station manufacturer's business. For a UMTS base station (WCDMA or TD-CDMA), or a CDMA2000 base station, the processes defined and controlled within this module cover physical layer and MAC layer. A base station vendor may have the entire device manufactured by subcontractors, but this part should always remain a core to the business and carefully safeguarded in-house. Algorithms for the control of the base station, resource allocation within a cell, call admission and control, handover, and many other processes are controlled by RNC software. The software needs to be updated regularly, and kept up-to-date with information on standards, new hardware, new technologies, and so on. This, of course, ensures a continuous, recurring revenue from these RNCs as earnings from software support and, from time-to-time, from hardware supply also, whenever upgrading becomes necessary. Such a manufacturer clearly holds significant control and bargaining power over the operator.

Typically, an RNC can control from tens to hundreds of base stations. They regulate and control some physical layer parameters between different base stations; they regulate and control resource allocation, handover of users, and so on. In case of handover, packet combining and decoding is also carried out at the RNC. And they also transfer information to and from gateways for interfacing the digital/analogue public network.

As IEEE standards only define layer 1 and layer 2 functionalities *(physical and data-link layers)*, IEEE 802.16 WiMAX base stations are expected to have a less complex control software. A central network controller does not exist in the present WLAN or WiMAX systems. There are, however, activities to specify and standardise this and higher nodes for future systems.

6.3.2 SGSN and GGSN

Serving GPRS Support Node (SGSN) switches packets from a user in the service area to and from the internet. Gateway GPRS Support Node (GGSN) is the gateway where a packet call is switched to the public internet and the other way around. These two nodes in the core network are used for the WCDMA standard packet-switched services. Several other nodes, designed for circuit-switched services exist: these are expected to be phased out as fully packet-switched services are used in broadband wireless systems. Similar functionalities as provided by these would be required in the future broadband wireless systems.

6.3.3 HLR

Home location register nodes store a user's service profile, including the location of the device and the services it can access.

6.3.4 Total costs

The cost of the network equipment is again confidential industry information, and is negotiated between vendors and operators. This author's estimate of base station-to-network equipment cost (RNC + other nodes) is in the order of 80% and to 20%, respectively, of the total infrastructure equipment cost.

Total Infrastructure Cost

It is difficult to estimate the cost of a network. Many operators provide a total figure as capital expenditure in their yearly financial reports. They do not generally disclose the detailed cost breakdown, nor do they provide the cost of equipment procurement. There is little information on how the infrastructure costs of an operator have been changing over recent years.

Assuredly, the cost has been falling. Almost all network equipment makers' revenue projections and share prices have been slashed since the burst of the IT bubble. The emergence of organisations such as OBSAI have increased the bargaining power of operators, encouraged competition, and driven prices down. The estimate given by three present and prospective operators in Japan, as listed in Table 6.6, may be a good guide on how prices

Table 6.6 Cost estimates for a 3G network roll-out in Japan

Operator	Starting in year	Estimated cost (US$)
DoCoMo	1999	20 B
Softbank	2007	2 B

have been moving. These numbers are this author's estimates for two operators for a full 3G WCDMA network roll-out, and the year the operator started their network build-up. These estimates point to how much infrastructure costs have been falling over recent years.

6.4 End-user Equipment

Several end-user equipment are shown in Figure 6.13. A mobile phone nowadays is also a data communications terminal. A laptop computer is also a data communication terminal, one among many applications that it supports. A wireless modem is a terminal with few, if any, application support, and generally provides wireless data connectivity (a bridge) to another device: for example, a computer.

Regardless of its type, the end-user equipment requires modules similar to that of a base station. Still, depending upon the end-user equipment functionalities, different components will be needed. A wireless modem has an ethernet output to connect a laptop to the internet using broadband wireless technology. Mobile phones, with browser functionalities have more sophisticated peripheral needs. For example, a video mobile phone needs to include

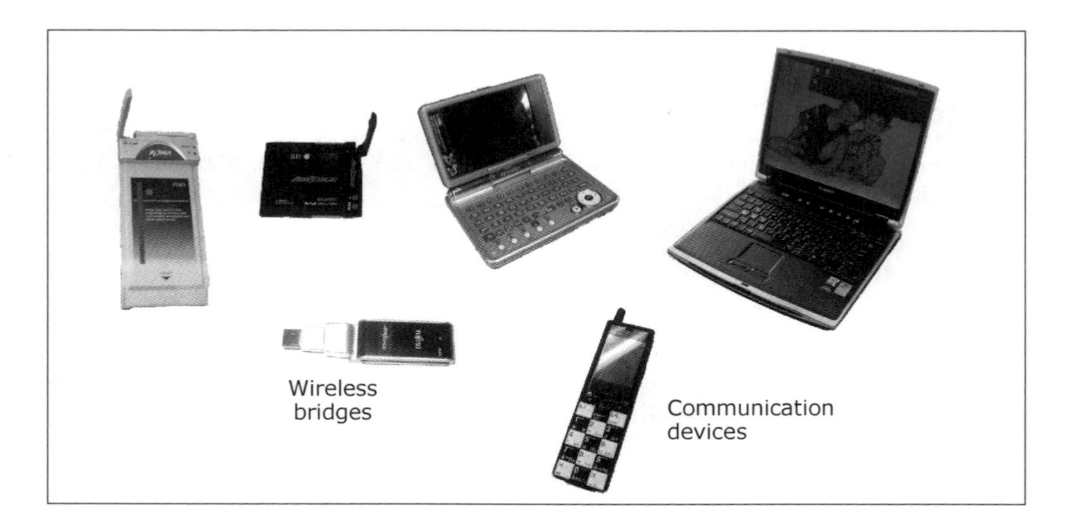

Figure 6.13 Some end-user equipment

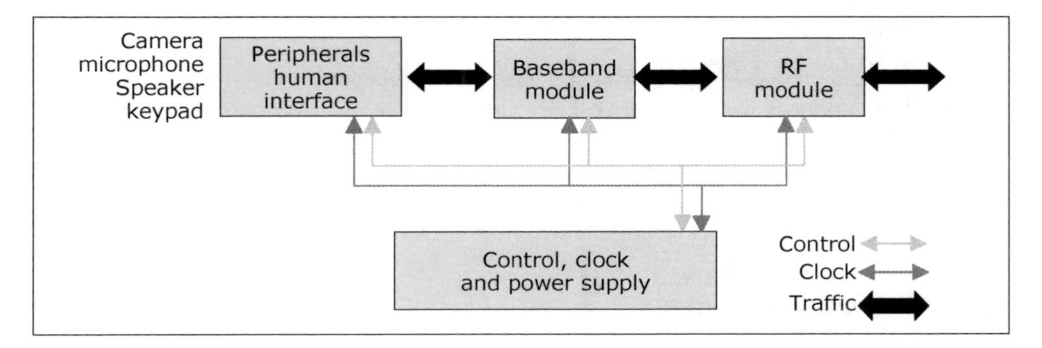

Figure 6.14 The structure of a user equipment device

a high-quality display as well as a digital video camera. The wireless communications unit in a laptop computer also acts as a bridge, connecting the applications of the laptop to the internet. Figure 6.14 shows the structure of end-user equipment with different modules.

6.4.1 RF module

The RF module in the end-user device acts in a way similar to that of the base station, but there are differences. The transmission power requirements are lower, leading to less expensive power amplifiers. The filtering requirements are not so severe as in the base stations. Antennas are also much smaller. And an end-user equipment communicates with only one device, meaning that multiple transceivers are not required. All of these distinctions lead to a much smaller cost.

Within the total RF cost, power amplifiers feature prominently. However, their smaller transmission power as compared with a base station means that they are much less costly. Here too, differences in input-back-off requirements can make a significant difference between CDMA and OFDM-based devices.

In user equipment power consumption, the maximum transmission power, and the degree of heat generation are important parameters. Mobile handsets have relatively smaller battery capacity, and lower maximum transmission power capability. They are also sensitive to heat generation, and need to use more efficient power amplifiers. Laptop computers have a higher device volume and battery power, but also a higher power consumption. At the same time, they are sensitive to heat generation, but not to the same extent as mobile handsets. Portable wireless modems can provide higher transmission powers and are more robust with regard to heat generation.

As discussed in Section 2.4.4, OFDM devices have a larger PAPR, and as a result, the power amplifier needs to operate at a lower transmission power level with less power amplifier efficiency. Naturally, this consumes more power and generates more heat. Such a condition is clearly not desirable for mobile handsets. Wireless modems, however, are more tolerant. This is one of the reasons why most pre-WiMAX customer premises equipment (CPE) devices have so far been wireless modems.

6.4.2 Antennas

The size of the antennas in end-user equipment clearly depends on the size and type of the device. Mobile phones have limited space for antennas, and usually contain two polarised antennas. If the end-user equipment is a laptop computer, then more antennas may be accommodated. The associated cost will vary accordingly. If MIMO antennas, or adaptive array antennas, are used, then corresponding RF control modules also need to be included, raising the cost further. Antenna cost, however, is a minor part of the total device cost.

6.4.3 Baseband module

The baseband module, too, is similar to the baseband module of the base station. As it is designed to connect to one device, the base station, it does not have multiple transceiver complexities. Again, advances in microprocessing power have facilitated the realisation of more complex reception algorithms, such as joint detection, in TD-CDMA.

6.4.4 Control software

The major difference between base station and end-user equipment, especially a handset, is with regard to the software. End-user equipment supports the provision of many applications. Peripheral functions related to browsing, display, security, and so on, need to be handled. The operating systems are quite complex. Table 6.7 shows some different applications that run on a typical 3G phone. These applications are associated with multiple data services that these phones access. Depending on the scope of a broadband wireless device, all of these functions may be required. The cost of this application software is generally higher than the hardware.

6.4.5 Clock and power supply module

Clock specifications for user equipment is less strict as compared to a base station. Power requirements, however, are strict for battery powered units. Batteries are a significant part of the units' costs, and their size determines significant parameters such as communication time (talk time) and stand-by time.

Table 6.7 An example of operating software within a DoCoMo 3G phone

Function and software	Made by
Security	McAfee
Browser	Access
Flash	Macromedia
Platform: Java	Aplix
Platform: BREW	Qualcomm
Operating system: Symbian	Panasonic mobile, others
Operating system: Linux	Panasonic mobile, NEC

Batteries power all of the electronics and processing devices in the end-user equipment. Furthermore, they power the transmissions made from the device in the uplink. The higher a technology's transmission power requirement, the larger the drag on the battery and the shorter the communication time. This could differentiate between the cost and size of devices using any of the broadband wireless technologies.

6.4.6 Peripherals

End-user equipment can have a number of peripheral devices. These range from keypads to displays, cameras, speakers, microphones and other devices that enable a human user to interface with the device – and through it, to interface with other equipment and with the public network. The mobile handset has emerged as a primary data communications terminal, which can connect to the internet. This has meant that many additional services can be delivered through the handset and that appropriate tools have had to be developed to better facilitate the mobile's use. As an example, the percentage of camera-equipped DoCoMo mobile phones has grown to about 75% from about 30% between June 2003 and September 2004. High-quality speakers and larger, clearer displays have also become necessary parts of mobile handsets to facilitate video and audio streaming services.

6.4.7 Total costs

Table 6.8 shows typical prices for several types of mobile phones in Japan. These are for mobile phones/devices, and wireless bridges. These are cost estimates after operator subsidies. While subsidies differ from operator to operator and the type of handset, these figures are indicative of how cost is distributed. We show an estimate of cost breakdown for a high-end 3G phone in Figure 6.15.

6.4.8 Device volume

The volume of the user equipment varies significantly according to the functions of the unit. Typical mobile phone handsets of 3G systems have a volume of around 100 cc, and a weight range of 70–150 g. The volume has changed significantly over the years. Figure 6.16 shows an analogue phone circa 1987, and a digital phone from 1998. The former has a volume of 500 cc and a weight of 700 g. The latter is 58 cc and 57 g. The sizes have grown since, with the addition of data communications functions and bigger displays, and cameras. The portable (as opposed to mobile) data user's equipment, such as pre-WiMAX devices

Table 6.8 Prices of some mobile phones and wireless cards

Parts	Cost (US$)
FOMA mobile	~$150
FOMA PCMCIA	~$100
Willcom flash	~$70
Willcom USB	~$70

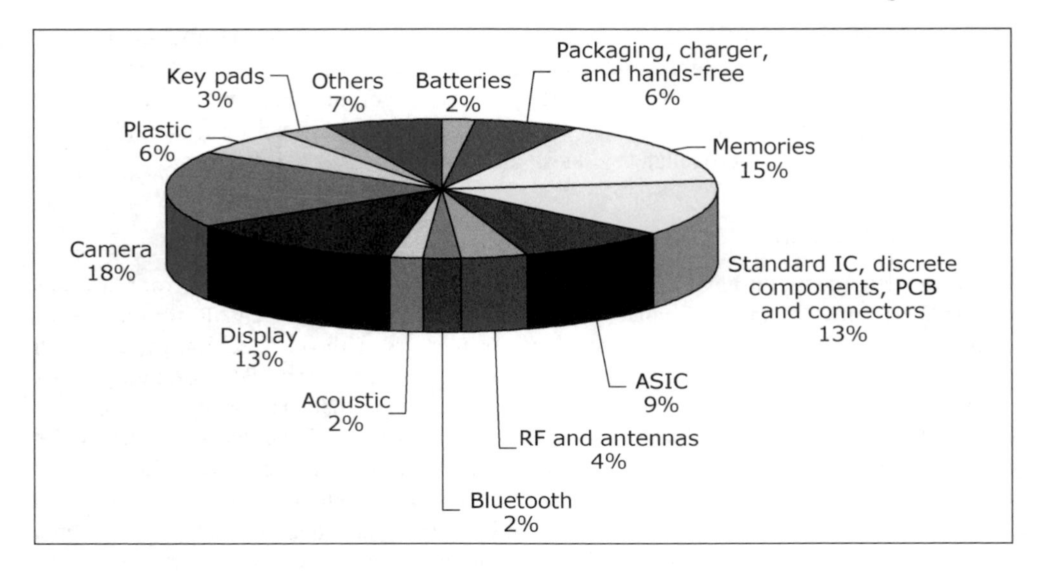

Figure 6.15 A typical costs breakdown for a mobile phone

Figure 6.16 A 1G and a 2G Panasonic mobile phone

associated with some IEEE 802.20 standards have much larger volumes in the order of 1000 cm^3. On the other hand, 802.11 wireless cards can have a sub-10 cm^3 volume. The transmission power required, as a function of coverage area, and the antenna size of the device, contribute largely to device size.

Summary

In this chapter, we have discussed the costs associated with the necessary equipment in a wireless communications system. Generally speaking, the advances made in processing technology and electronics have meant that the cost of building equipment has significantly fallen. Base station equipment cost as little as one-tenth of what they did five to ten years ago. Equipment sizes have also shrunk. Costs of future broadband wireless communication systems are expected to follow the same trend.

Although the cost trend has been downwards, it has not exhibited the same rate for all modules. While signal processing has been following the Moore's law, that is, doubling in speed every 18 months, electronics associated with filters and amplifier devices have not exhibited the same refinement. This means that the cost of those technologies that rely on improved signal processing will decrease faster than those that need better amplifiers for example. This is a significant difference between CDMA-based systems, and OFDM-based systems, the former requiring faster processing, and the latter better amplifiers.

As for end-user equipment, here too the costs have been falling. Addition of more advanced accessories has kept the retail prices steady.

Further Reading

- OBSAI web site: `http://www.obsai.org/`
- On Wireless systems and components:
  - Akaiwa, Y., *Introduction to Digital Mobile Communications*, John Wiley & Sons, 1997.
- On Processing requirements:
  - 3GPP report TSG-RAN-1 Meeting #35, R1 031 301
- On Power amplifiers:
  - Venkataramani, M., *Efficiency Improvement of WCDMA Base Station Transmitters Using Class-F Power Amplifiers*, Master's Thesis, Virginia Polytechnic Institute and State University.
  - Cripps, S. C. *RF Power Amplifiers for Wireless Communications*, Artech House Publishers, 1999.

7

Network Design and Operation

To a large extent, mobile communications is a fixed-cost business as we have discussed. The first major element of cost is acquiring frequency spectrum for operation, as was discussed in Chapter 5. The second major element is associated with the purchase of network equipment, as was discussed in Chapter 6. In this chapter, we discuss costs associated with the design of network, its roll-out, and its maintenance and operation. These costs are partly fixed and partly variable. We will also discuss costs associated with running a network and serving customers: costs that are by and large variable in nature.

There are several aspects to the network design and planning stage of the business. The operator must decide on which markets to target; how broadly to cover a country/region; where to place the base stations or access points; which network topology to use and how to connect the base stations or access points to a core network; how to facilitate further evolution of the network's design; and so on.

After the network is designed and rolled out, and service is introduced, the network must be maintained. The operator needs to regularly test the network to ensure that a high quality of service is maintained. As the customer base grows, the operator needs to improve the service, revise its network plan, add new cell sites, and retune the network. These costs can be classified as variable, since they depend to a great degree on the size of the customer base. Furthermore, there exist variable running costs such as spectrum fees, power consumption, site lease and rental, human resources, and so on. Alongside these are more variable costs such as those associated with customer acquisition and service, billing, and end-user equipment subsidies.

Again here, we demonstrate how new technological developments can contribute towards reducing the costs. It is shown that technologies can be utilised to replace some fixed costs with variable costs, which can reduce the initial capital expenditure.

7.1 Network Design and Planning

The reach and coverage of a broadband wireless network should be a business decision made by the network's operator. At times, this may be possible. However, in many cases,

Broadband Wireless Communications Business: An Introduction to the Costs and Benefits of New Technologies Riaz Esmailzadeh

the providing of a (nearly) universal wireless access coverage is a contractual obligation placed on the operator by the government. The cost of universal coverage is high; many areas can be sparsely populated, and the revenue per base station in these particular locations can be less than the cost of service provision.

Regardless of its contractual obligations, an operator may still decide to provide wide coverage. This ensures the provision of high grade of service and wide-area user connectivity. Indeed, the extent of coverage is primarily a business decision; lack of widespread coverage has been a primary cause of failure for a number of operations. A primary example of this is the personal handy phone system (PHS) service in Japan. While the service was initially successful when introduced in 1994, the operators have not flourished. One main reason has been its inferior coverage compared with other mobile technologies (see Box: Personal Handy Phone System). The advantages of wide coverage can be greatly offset by a higher cost of infrastructure.

Personal Handyphone System (PHS)

A main reason for the relative lack of success of the PHS standard in Japan is the fact that the service coverage was not wide enough. Despite offering the lightest handsets with long standby and talk times, and better voice quality, the original three PHS operators have struggled competing with other operators:

- NTT Personal Communication was absorbed by DoCoMo. The PHS service will be discontinued after 2007.
- ASTEL ceased operation in 2004.
- DDI Pocket was acquired by the Carlyle Group, renamed Willcomm, and is the only ongoing operator.

Despite its record in Japan, PHS services are doing well in China.

Technology helps to ensure that a coverage as large as possible is achieved with a fixed number of network equipment. This is the art of network design. Network design and planning helps to achieve a desired level of coverage over selected areas using the minimum number of base stations and other network nodes.

Network design and planning is the business of deciding how many, and what type of, base stations are required for a certain populated area, and where these base stations should be located. Network planning is a major part of the initial capital expenditure. It is estimated that the network planning services and the required software tools' costs make up about 25% of the total infrastructure costs.

7.1.1 Stages in network design

There are several stages in designing a network. First, an estimate is made of the total expected traffic, on the basis of which the number and type of base stations required are determined. After this, cells are designed, cell sites are secured, and base stations and antennas are installed. Next, the network is 'tuned': the antennas, output power, and other configurations are set in such a way as to ensure maximum coverage and throughput.

This is also referred to as network optimisation, and includes time-consuming drive tests throughout the coverage area to ensure that optimal coverage is achieved. After this, base station traffic statistics, as well as further drive tests are used to upgrade network design.

Traffic engineering, as discussed in Section 2.8.2, is an important technology in network design. Generally, each base station can serve a finite number of users, and process a certain amount of offered traffic. Therefore, the number of base stations required over a coverage area is a function of the expected number of users and traffic, which is itself a function of total population. A sparsely populated area of, for example, 30 km^2 kilometres, may be served by a single base station. However, an area of the same size but with a higher population density may need tens of base stations to provide a desired coverage and grade of service. Traffic engineering is the first step in estimating the traffic and required size of transmission resources (channels). From this, the number of base stations required may be calculated.

In the early days of mobile communications, cell planning was carried out by a subsidiary of the operator. With the explosive growth of mobile communications in the mid- to late-90s, many specialist network design companies entered the market. Because network design and planning services, together with fine tuning and maintenance, are a time-consuming and human-resource intensive job, many operators now have this service outsourced.

Cell design tasks are summarised in a value chain in Figure 7.1. Site acquisition and engineering are usually carried out in parallel with network design and planning. Installation of the base station is then carried out, followed by network optimisation, and regular maintenance. As stated above, some operators perform all of these processes in-house, or sometimes through a fully owned subsidiary. Different degrees of outsourcing are also possible. Naturally, a cell planning company would like to carry out as many of the activities of the value chain as possible.

The cost of network design, as was also stated in the previous chapter, is very much a function of the bargaining power. Network design companies' fortunes have followed the fortunes of infrastructure vendors: high demands led to good business and high valuations in the late 1990s. As demand subsided, and as more competitors entered the market, margins and stock valuations have sharply declined, a trend accentuated by the general downturn in the IT industry after the year 2000.

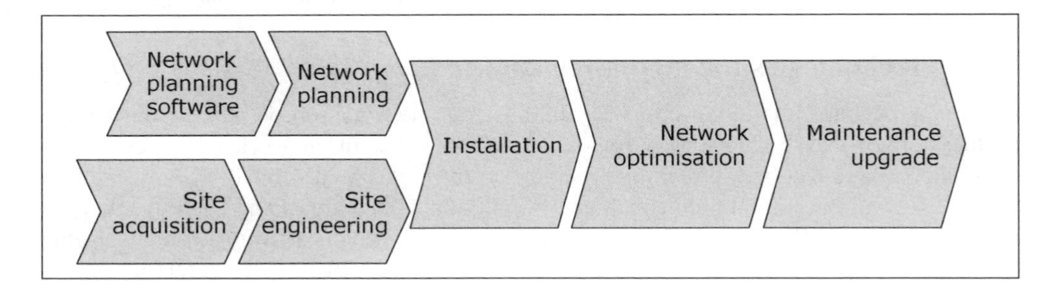

Figure 7.1 Cell design value chain

Two examples of these design companies are shown in Box 'Network design companies'.

Network design companies
Wireless Facilities Inc. (WFI) is one of the world's leading network design companies. They offer a range of services, from cell planning to turnkey solutions. WFI generally use third party vendor cell planning design tools. Aircom are mostly known for their cell planning software: although they offer network design services too.

7.1.1.1 Different technologies

With largely similar frequency utilisation efficiencies, the UMTS CDMA-based and the IEEE 802.16 WiMAX technologies can be expected to require a similar number of base stations for networks with similar coverage parameters. Using different technologies in each standard may lead to different network designs. However, technologies do not seem to be as significant to cost as do operator strategy issues: what degree of coverage and what particular market segment is the operator targeting?

Although frequency utilisation efficiencies may be similar, network capacity, in itself, is a function of total allocated spectrum. A WCDMA base station with 2×20 MHz of allocated bandwidth has a higher traffic capacity than a TD-CDMA base station with 1×15 MHz of allocated bandwidth. A comparison between CDMA-based and OFDM-based technologies follows the same argument. From the point of view of a network designer, installation of a base station incurs one unit of cost. Therefore, the greater the user capacity of a system, the fewer would be the number of base stations required. This is regardless of whether the increased capacity is as a result of larger bandwidth, advanced processing, or the use of adaptive antennas.

Network design for 802.11-based systems follows a different argument. These systems are largely designed without taking into consideration the size of the area, or universality of coverage. Although a number of companies design for 802.11-based networks, the cost is not so high. Issues of concern here are also different. Where an 802.11 access point may be placed in a McDonald's outlet is equally a question of aesthetics as coverage.

7.1.2 Technologies for increasing capacity

For a new operator, the initial cost of a full-coverage network roll-out can be quite high. Furthermore, it may be logistically impossible to roll out a full network. With zero initial subscribers, there can be a prohibitively high risk for high initial capital expenditure. It is desirable to roll-out a minimum coverage of highly populated areas, and subsequently 'fill in' the network as offered traffic increases and the network becomes saturated and incapable of adequately serving the need. To increase system capacity, one way is to introduce new base stations within the network to service the increased traffic.

Introducing new base stations, however, is not an easy undertaking. Figure 7.2 shows how two neighbouring cell antennas are tuned. As the signal from a base station appears

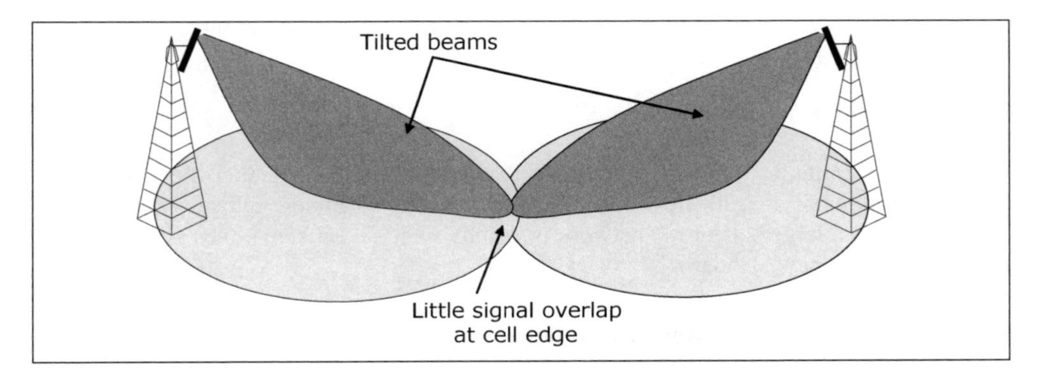

Figure 7.2 Antenna tilting to optimise coverage

as an interference to the neighbouring cell's users, the antennas of base stations are tilted in order to minimise the mutual interference. Adjusting the antenna tilts is one lever in controlling and tuning the network.

Should a new base station be introduced now, all antenna tunings must be redone. Moreover, there is no suitable place to place the base station and create a cell around it, as can be seen from Figure 7.3. The networks have therefore been traditionally designed with long-term traffic forecasts. Furthermore, to increase system capacity, the following technologies are used rather than adding new base stations:

- Introducing, or increasing the number of sector antennas; this can increase system capacity several fold.
- Usage of adaptive array antennas, similar to the above.
- More advanced transceivers at the base station (useful for 2G and beyond). Using more advanced signal processing techniques.

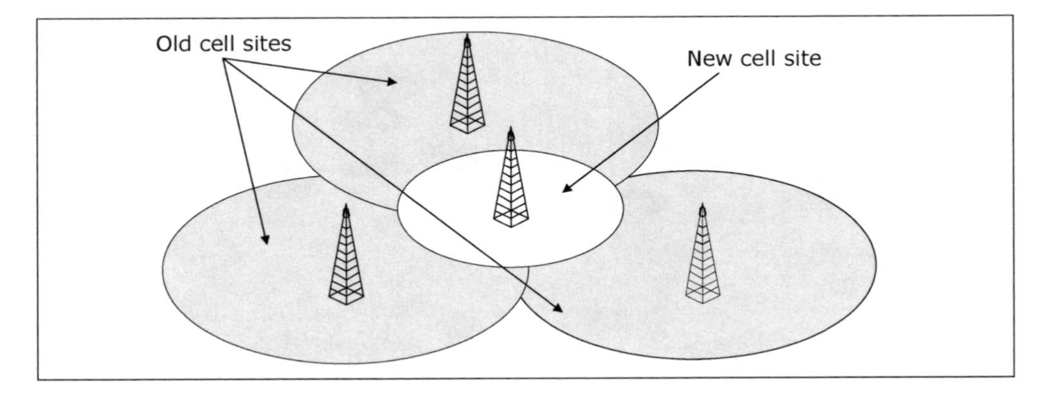

Figure 7.3 Introduction of a new base station

- Reducing frequency reuse factor, and through it, increasing the number of frequency carriers used in one cell

- Acquiring more spectrum

There are limitations, however, to all of these solutions, but together they can ensure that introduction of new cell sites is significantly delayed. When new cell sites are required, the redesign of a large part of the network is usually undertaken, sometimes along with the introduction of a new standard.

7.1.2.1 Hierarchical and overlay cell design

A new operator, nevertheless, benefits from a low initial capital expenditure. It is desirable to design a network with a very wide area of coverage in order to ensure wide service availability, and as the number of subscribers and traffic increases, more cell sites may be deployed in areas of large traffic concentration. However, the solutions listed in the previous section may not work for this operator, as it has started with a bare minimum number of macro-cell base stations. One solution is to use an 'overlay' structure, as illustrated in Figure 7.4.

We have already seen in Section 4.1 that different distributions of population density and traffic needs lead to cell designs with various degrees of cell radius. A mixture of macro-, micro- and pico-cells are required to design a network with adequate coverage and user capacity. These structures should not, however, overlap, as mutual interference can result in reduced overall throughput. The overlay structure must find a solution to reduce mutual interference

Overlay structure has been researched for over a decade for FDD-based CDMA systems. It has been realised, however, only in areas where the two cell structures are largely mutually exclusive. For example, a micro-cell is deployed under a macro-cell but only in the subway station. The reason is that if the two systems operate in overlapping areas, then they become mutually interfering. This leads to degradation of system performance.

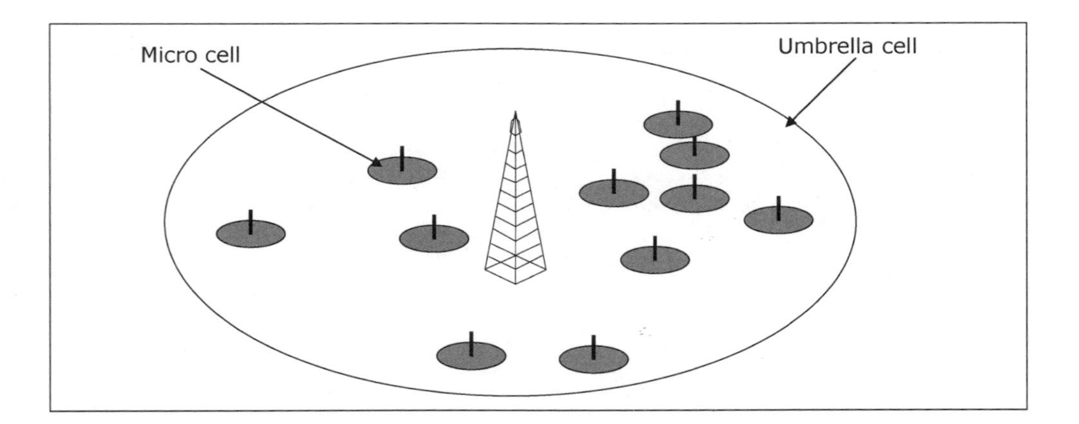

Figure 7.4 Umbrella cell structure

One technology that enables overlay system design depends upon the choice of duplex mode (see Section 1.5). While overlay structures do not work well in FDD systems, they can be made to operate more easily in TDD systems. This is particularly true with the TD-CDMA standard, as its slot structure, and its flexibility in allocating transmission resources, enables a fully overlapping overlay structure. This facilitates the introduction of a broad coverage system with the gradual expansion of capacity. We had discussed this technology in Section 4.4.

7.2 Site Cost

Cell site selection and preparation is another stage of network roll-out. The cell planning process identifies where cells should be located. The selection of a physical place as a cell site can be quite complex. Factors to be considered include:

- Height of the building/tower: Antenna height of a macro-cell base station is usually between 30 to 50 m from the ground. Therefore, the base station/antenna would be installed either at the top of a tower specially built for it, or on a sufficiently high rooftop, as illustrated in Figure 4.4.

- Proximity to other networks' sites: Any two systems operating at neighbouring frequency bands interfere with each other. This places restrictions on how close, for example, antennas of a TD-CDMA and a WCDMA systems can be placed. Generally, antennas need to be adequately apart, and extra-special filters installed if the cells are co-located or are too close to each other. Furthermore, sectorised antennas should be directed away from each other.

- Fixed network connectivity: A preferred solution for connecting a base station to an RNC is by high-speed optical fibre. Availability of such connections at the site, or ease of access to optical fibre connections is advantageous.

- Power availability: A base station requires significant power supply. This is of particular concern in rural installations, and small power generators may be required.

- Site reliability: The building or tower needs to be structurally sound.

- Environmental concerns and aesthetics: Civil rules and regulations may restrict the size and height of antennas that are to be installed. And large array antennas, for example, may not be acceptable.

7.2.1 Rooftops and towers

The associated typical construction and rental costs in Tokyo are listed in Table 7.1. As Tokyo rental costs are amongst the highest in the world, relevant adjustments should be made for other cities. However the high density of multi-story buildings in Tokyo reduces the requirement for tower construction. Also included in the table is cost associated with building a tower for antenna placement.

Table 7.1 Typical costs for a cell site

Item	Typical cost ($)
Site rental	1000–5000 per month
Site preparation	50 000–100 000
Tower construction	100 000–300 000

7.2.2 Micro/pico base station sites

Smaller size base stations, such as those for micro- and pico-cells have smaller antennas and power requirements. The antennas are usually installed at much lower heights of 10 to 15 m – on an electricity pole, for example. Costs of installing micro-cells, including site rentals, are significantly less than for macro-cells.

7.2.3 Installation

The cost of installation can be divided accordingly as mechanical, constructional, and electrical. With the size of base stations (see Section 6.1.6) decreasing, the costs associated with installation have also decreased. The major construction costs however, are associated with antennas. As adaptive array antennas are larger than sector antennas, the associated cost is also more. At higher operating frequencies, though, antenna size decreases, and so does its installation cost. These costs very much depend on the base station site engineering issues: if special preparation is required, the cost naturally increases.

Older-generation base stations were usually housed in a special air-conditioned room, which could be separated from the antennas by tens of meters. This added to the cost of installation, because special cables were required to connect the base station to the antenna, and also the air-conditioners had to be put in place. With the smaller size of base stations now becoming possible, and less stringent air-conditioning requirements, base stations may be placed in a cabinet on rooftops near the antennas, thus reducing the cost of installation considerably.

7.2.4 Tuning

When a base station is installed, and its antennas are connected, then the system needs to be tuned. This mainly involves adjusting and tilting the antennas, particularly sector antennas, to ensure that the area is well covered, that there are no dead spots, and that neighbouring base stations do not interfere. Usually, drive tests are carried out throughout the coverage area to measure reception quality. This is quite a time-consuming exercise because of the need for many trials and adjustments.

7.2.5 Maintenance

Base stations need to be upgraded regularly. Different software releases need to be installed, and hardware needs to be updated, replaced, or otherwise maintained. This, again is a major fixed cost, and depends on the number of base stations.

Maintenance is a labour-intensive cost. Some operators have this outsourced to network design and engineering companies, and some rely on their own subsidiaries.

The newer design of base stations allows for software upgrades to be done remotely. The newer hardware is also designed in such a way as to reduce the need for replacement. An example is the IPWireless' TD-CDMA base station, which can be updated remotely, using software upgrades.

Regardless of technology, base station hardware needs to be replaced regularly, even though the frequency of replacement may decrease. Larger processing power and newer releases of a standard make older hardware obsolete. What the newer designs of base stations can aim at achieving is a lengthening of the period between upgrades. Older 2G and 3G systems are estimated to require software and hardware upgrades almost once a year. If this frequency can be reduced by 50%, the maintenance cost will be accordingly reduced.

7.3 Backbone Fixed Connection

Base stations connect to the fixed network and to the public internet via RNCs. The connections, as discussed above, need to be very fast: required fixed-link transmission rates correspond to the data transmission rates of broadband wireless systems and are in the order of hundreds of Mbps.

Generally, fixed connections may be wired or wireless. Wired connections are usually optical fibres. These links are widespread over urban areas and have become quite inexpensive. In suburban and rural areas, optical fibres' availability is not as common. When optical fibre links are not available, point-to-point wireless links are used to connect a base station to its RNC. A number of wireless link devices exist with different possible bit rates. Typical costs for optical fibre lease in the Tokyo area and fixed point-to-point devices are shown in Table 7.2.

The transport mechanism for these connections varies from technology to technology and the cost is accordingly different. In older 1G, 2G, and 3G systems, asynchronous transfer mode (ATM) devices were used to connect base stations to their RNC and to other network nodes. Connection to the public telephone network was also through ATM.

With the emergence of IP as the preferred mode of transport, and the availability and scalability of IP switches, it is expected that all future broadband systems will be all-IP. Broadband wireless systems are expected to be all-IP, as discussed in Chapter 3.

7.3.1 Physical connections

Extensive fibre networks have been developed over the past decade in many developed countries. In Japan, several companies have optical fibre network and fibre to the home

Table 7.2 Cost of fixed backbone links in Japan

	Typical cost ($)
Optical fibre lease	1000–4000 per month
Wireless point-to-point	30 000–50 000

(FTTH) services are quite widespread. Public communications networks reach to within very short distances of customer's locations in view of the high population density. Also, the public train transport companies and utility companies have laid in their own optical networks. Primarily, these have been for the purpose of controlling these companies' own services. However, their surplus bandwidth is offered for communication services (to operators who do not have a network of their own, for example, a new operator). As a result, a large percentage of base station sites can be connected to their RNCs and to the backbone using the surplus bandwidth of these optical fibres. Where these are not available, fixed wireless links are required. DSL connections using twisted copper wires are another option, but in the majority of cases, where high DSL rates are possible, optical links also do exist.

This may not be the case in other countries. Lower population densities increase the cost of fibre to the home, or to the curb. Furthermore, DSL services cannot achieve the ideally high transmission rates because of the greater average distances between residences and the exchange. These factors affect the cost of network roll-out and backbone connections, leading to other networking configurations that mainly use wireless technologies.

Connections between a base station and its RNC may be made using DSL modems over twisted-pair copper media. However, as these have a limited throughput, especially in the uplink, their use is limited. These connections can be considered for mini-base station configuration as discussed above. They can also be used for connecting WLAN to the internet where optical fibre is not available.

7.4 Networking Based on Access Points

The cost of networking for IEEE 802.11-based structures is somewhat different because of their smaller coverage areas. The relatively larger number of network nodes leads to higher costs in relations to:

- Greater number of fixed links for connecting access points to the backbone network.
- Maintenance costs: more sites to visit (although maintenance cost per site may be less).
- Installation cost: more sites (although cost per site may be less).

Furthermore, service management issues arise also in relation with site suppliers. Since access points are placed on other businesses' premises, the legal and contractual concerns are different.

7.4.1 Access points and antennas

WLAN access points and the antennas for 802.11 devices are accommodated in the same unit and therefore the installation is simpler. Neither is there a need to tune the antenna. Still, network design is required for the coverage area, and cell design can be different between the enclosed areas and the open areas. A detailed cell design to reduce the inter-access-point interference may be futile as operations are done within an unlicensed band. Other operators may very well exist in the same area.

As stated earlier, aesthetic considerations can be critical where access points are installed. Access points need to be connected to the backbone network. These can be DSL links with a 'best-effort' few Mbps capacity, or optical fibre links with generous capacity.

Estimates for the cost of an access point with installation charges, including aesthetic considerations, can easily run as high as five thousand per site.

7.4.1.1 Adaptive array antennas

There are a number of studies on extending the coverage of IEEE 802.11 devices to a kilometre or more using adaptive array antennas. Should technical challenges for such a system be solved, and these extended systems be realised, installation charges similar to a typical WCDMA base station, for example, can be expected.

7.4.2 Maintenance

Maintenance costs for access points can quickly add up as the number of these devices are considerable. Smaller device sizing, and their lowered cost means that it becomes economical to replace access points regularly with upgrades.

7.4.3 Fixed network connection

Each WLAN access point needs to be connected to the internet using fixed links. In a manner similar to macro-cell base stations, the connection may be optical fibre, DSL, or wireless point to point. Again, similar lease cost as above can be expected.

WLAN backbone wireless point-to-point connections may be based on IEEE 802.16 technology. Indeed, the high cost of WLAN backbone connectivity was an initial purpose of the 802.16 standards development.

7.5 Customer Costs

Another major cost of wireless communication service business is that associated with end-user management. There are costs related to initial customer acquisitions, followed by costs for customer service and maintenance. These costs are generally variable: they increase as the number of customers increase. This section gives a brief description of these costs. Here, new technologies are not of great benefit in reducing these costs.

7.5.1 Customer acquisition

Customer acquisition is perhaps the biggest initial variable cost. This cost is naturally a function of supply and demand. It was high during the initial introductory stages of mobile service. During the boom years, as more people wanted mobile phones, the cost of customer acquisition fell. Again, demand fell as the market became saturated. Competing operators now aim at stealing customers from each other, again raising the acquisition cost. Regulators have somewhat reduced this cost by allowing number portability between operators (see Box: Number portability).

Number portability

Number portability allows users to change their mobile operator while keeping their old mobile phone number. Here is a list of some countries where number portability is available, and the year it came into effect:

1997	Singapore
1998	United Kingdom
1999	Hong Kong, The Netherlands
2000	Spain, Sweden, Switzerland
2001	Australia, Denmark, Italy
2002	Belgium, Germany
2003	France, Austria, Portugal
2004	USA, South Korea
2006	Japan (expected)

Initial acquisition costs include advertising (a fixed cost) and subsidies (a variable cost). While advertising is used both for retaining present customers and for attracting new ones, subsidies are mainly aimed at attracting new customers.

7.5.1.1 Subsidies

Subsidising initial customers is a normal practice at the service introduction stage. Typically, a part of the cost of the mobile device is subsidised. Sometimes, subscription plans are available with handsets offered at zero cost to the end-user. These plans usually call for a continuous subscription of 12 to 24 months. In Japan, handsets are offered to first-time subscribers at very low cost, without the obligatory subscription period. Such subsidies can initially cost an operator $400 or more in Japan. Mobile phone subsidies have existed in Japan over both the periods of low and high growth. In other countries, where the growth rate is high, subsidies can be significantly less, or none at all.

7.5.1.2 Subsidies for broadband services

Similar subsidies can be expected for broadband wireless end-user equipment. This is because similar subsidies are being offered for DSL customer premises equipment (CPE)s. As the DSL services are at the initial service introduction stage, the subsidies were deemed necessary for encouraging growth (see Box: ADSL customer acquisition). The use of customer acquisition subsidy for broadband services will also depend on the cost of broadband wireless end-user equipment. The initial equipment will possibly have fewer peripheral devices, such as cameras or displays. The cost therefore, is lower, and the total subsidy can be lower. An estimate for these subsidies can be $200 or less per customer. Here, the cost of customer acquisition is more a function of finding and broadening the customer base with new services, which cannot be provided as easily with present technologies.

ADSL customer acquisition

Competition in the Japanese ADSL market has been fierce, and has resulted in making Japanese ADSL service the least expensive in the world. Estimates for customer acquisition is around $600 per new subscriber. A new subscription usually comes with free modems, and two to three months of free ADSL and IP telephony.

7.5.2 Customer service and billing

Customer service is yet another variable cost. The size of the customer service department is proportional to the number of subscribers to the service.

If broadband services were offered at a flat rate, then it could be expected that customer service requirements, particularly with regard to billing, would be reduced. These costs may not be very significant to an established operator with a billing system in place. However, it may be of great interest to a new operator, as a means of reducing the complexity of its billing system, and therefore the cost of establishing its new business.

Content providers, whose services are carried over an operator network, can also be considered as the operator's customers. Service and accounting issues related with these services also need to be considered in setting up the customer services policy and support department.

7.6 Other Operating Costs

We now briefly talk about some of the other operating costs: power consumption, frequency fees, and human resources.

7.6.1 Power consumption

Electricity is a major operating cost. Power is used for reception and transmission signal processing, as well as for amplification and actual signal transmission.

Power is also consumed for cooling base stations. Installing smaller, outdoor base station can reduce an operator's electricity bill significantly.

7.6.2 Spectrum fees

An operator may be charged a nominal fee for spectrum use. The fee may be structured on 'per subscriber' basis, as, for example, is the case in Japan. Or it may be a flat fee per MHz of spectrum, that is to say, the cost is either variable or fixed. Japanese spectrum fees are around $2. These are the fees for mobile phone services. How broadband wireless spectrum's use is charged remains to be seen.

Table 7.3 Numbers of employees for various operators

Operator	Business	Number of employees	Number of subscribers (million)
DoCoMo	Mobile operator	5856	(Mobile) ~50
KDDI	Mobile/fixed-line/ADSL operator	~8000	(Mobile) ~18
eAccess	Wholesale/enterprise ADSL operator	700	(ADSL) ~1.9

7.6.3 Human resources

Yet another major cost of the operator is that of human resources. The number of employees for the three major Japanese operators are listed in Table 7.3, next to their respective number of subscribers. Included also is the same for a wholesale Asynchronous Digital Subscriber Link (ADSL) provider, eAccess. These numbers are indicative of how many functions of the operator are outsourced.

7.7 Wholesale Operators – Network-less Operators

Finally, a word on wholesale operators and network-less operators. As the name suggests, a wholesale operator concentrates on operating a wireless network and does not manage end-users. An operator without a wireless network is the complementary entity. These operators manage the end-users, incur the cost of customer acquisition and management, and establish the brand. While a pure wholesaler is a rare breed, many network-less operators do exist. Many operators use their network to operate both under their own brand and also to provide a "pipe" for network-less operators. Services provided by the same physical network through the two operators may be very different, reflecting each operator's service strategy, the target market, and the content provider alliance. See Box 'MVNOs'.

MVNOs

Mobile virtual network operators offer communications services using other operators' network. These operators concentrate on brand development and customer service. Two examples of such operators are Virgin Mobile and TELE2.

The situation is similar in the fixed, broadband data service domain. There exist 'pure' wholesalers. These may be operators who own, for example, an optical fibre network, but the provision of service is clearly outside their area of business. Railway companies in Japan are an example. There exist also network-less operators who buy wholesale and manage end-user customers.

It is entirely possible that a pure-wholesale broadband wireless operator may emerge, especially in saturated markets where new customer acquisition is difficult. A wholesale operator may sell wireless services to current subscribers of a fixed broadband network. It may even sell to other wireless operators who are short of capacity, or who have a legacy 2G/3G network but no high-speed data communications network.

Summary

In this chapter, we have discussed on aspects of broadband wireless communications that have both a fixed and a variable cost behaviour. We first discussed wireless network design issues and costs. Traditionally, these costs have been mainly fixed: an initial major capital expenditure, followed by relatively lower costs for maintenance. New technologies, such as advanced antennas and receiver can reduce the initial outlay. Furthermore, new network topologies can facilitate cost structures that are more variable than fixed. Other costs are associated with leasing backbone fixed links, which can have both a fixed and a variable component. We then discussed costs associated with customer acquisition, service, and maintenance. These are mostly variable costs, and increase as customers increase. New wireless technologies can lead to a profitable flat-rate operation and result in a decrease in costs associated with customer service.

Further Reading

- On Traffic engineering:
  - Kleinrock, L., *Queuing Systems*, John Wiley & Sons, 1975.
  - Zander, J., *Radio Resource Management for Wireless Networks*, Artech House Publishers, 2001.
- On Network planning:
  - Laiho, J., et al. (Editors), *Radio Network Planning and Optimisation for UMTS*, John Wiley & Sons, 2002.

8

Services

Higher bit rates and full packet switching are the major technological characteristics of future broadband wireless communication systems. Still fully unknown is what the services that will be provided over these technologies are going to be.

The expression 'killer app' has been used in referring to a single major application that will become extremely popular and will generate large amounts of revenue for mobile communication operators. The term was coined in particular for the new data services of 2.5G and 3G systems. While universally popular killer apps have failed to appear, there are a variety of services that have appeared with limited popularity. While some have a degree of universal following, many of these services, in reality, have a niche market. What is important is that the combined revenue of all these services has been growing rapidly. The killer app now appears to be the composite of numerous nonuniversally popular services.

With the arrival of much higher data rates, and a full IP backbone system, the range of possible services widens significantly. Almost any service that can be delivered via a fixed network to a personal computer, is conceivably deliverable (and marketable) to a mobile subscriber. Indeed, as we have discussed before, one major promise of broadband wireless systems is that they enable wireless access to the internet for any device. From a practical point of view, the shape and form of the new mobile devices is the principal differentiating factor between mobile and fixed device services.

Several particular characteristics of mobile devices help give rise to various classes of different services:

- Convenience and simplicity: Mobile user interface is simple; there are no complex start-up and access procedures. This makes the providing of service to a larger population segment possible, including those who are less technically savvy.

- Relatively short access delay: There is no boot-up waiting time since mobile devices can be expected to be continuously online. This gives rise to interactive services, such as (impulsive?) betting.

- Personal nature of mobile devices: A mobile is associated with a person. This means that customised content – advertising and other information – can be delivered directly

Broadband Wireless Communications Business: An Introduction to the Costs and Benefits of New Technologies Riaz Esmailzadeh

to the user. Also, security information such as online banking passwords can be stored with more confidence.

- Location issues: There are two aspects associated with location. First is the mobility, which enables service access from any location. Secondly, the network can know, to within a certain accuracy, where the user is located. These two in combination lead to services related to the user's location.

On the tariff front, services with richer content and, correspondingly, larger traffic volume, stand a better chance of becoming popular with customers if the operators provide flat-rate charging plans. Flat rate is an industry trend which is expected to facilitate the spawning of many new mobile services.

Form factor

The services an end-user subscribes to are very much a function of the form of the end-user's devices. Several portable devices are shown in Figure 8.1. A laptop PC can be defined as being more portable than mobile. Its main function is not necessarily communications. However, as a communications device, its uses range from sending and receiving files, to web browsing, to email functionality, and to carrying out a video conference. On the other end of the spectrum is a mobile phone, which, as a communications device, has capabilities similar to that of a laptop PC. The capabilities are limited, however. On the other hand, the mobile telephone is significantly more 'mobile' than a laptop PC. Therefore, the broadband mobile communication services targeted for the users of these two devices will necessarily differ.

In this chapter, the services which may be provided over future broadband wireless systems will be discussed. In order to learn how these services can be provided over the wireless technologies discussed in this book, we classify these services by their quality of service requirements according to a model designed by the UMTS Forum, a telecommunications industry association of operators, manufacturers, and regulators. The classification is based on a user's perspective of the different services required. We also compare these services on their required transmission rates, total on-air requirements, and their delay constraints.

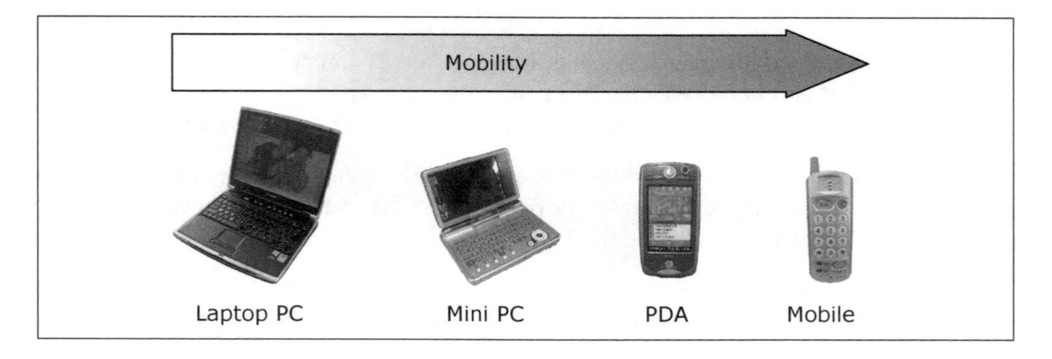

Figure 8.1 Some end-user devices

Furthermore, we discuss how these services can be billed; that is, how these services can generate revenues for both the content providers as well as the network operators.

8.1 Revenue Flow

In the early days of mobile telephony services, the revenue flow was simple. As illustrated in Figure 8.2, revenue flow from the subscriber to the operator was a simple process. This picture changed with the emergence of prepaid services, mobile virtual network operators (MVNO)s (Section 7.7), and has been further changing with the emergence of data services.

A general revenue flow diagram may be drawn as in Figure 8.3. Here, revenue flows from customers to service providers and then further into third-party content providers. The service provider also receives revenue from content providers. Revenues may also flow directly to the third parties, and from there, indirectly to the operator. In the following section, we will define different classes of services, and based upon these service definitions and operator business approaches, we will give more detailed examples of how these revenue flows may appear.

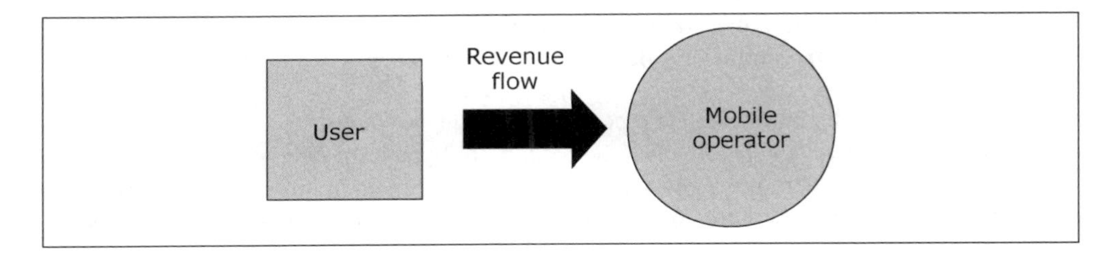

Figure 8.2 Revenue flow in mobile telephony

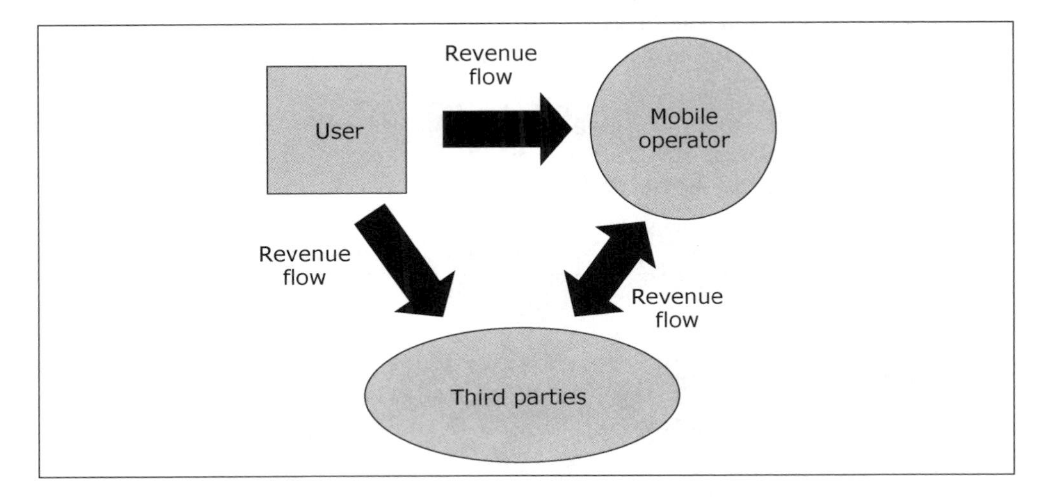

Figure 8.3 General revenue flow diagram for new services

8.2 Value Chain

The value chain for mobile services is complex, with many components and sub-components. One value-chain definition, for example, is suggested by Paavilainen (see 'Further Reading' list at the end of this chapter). This model considers the value chain with a focus on mobile communications business (m-commerce) elements, and divides the value chain into various elements for mobile commerce business as follows:

- Network: Includes all network players such as operators, manufacturers, and back-haul operators.
- Mobile Commerce: Includes players who develop software solutions for security, payment, and user-applications.
- Content: Includes content providers and content aggregators.
- Interface: Includes end-user device manufacturers and portal developers.

A UMTS Forum report provides a content delivery value-chain diagram as perceived by an end-user. The diagram is reproduced in Figure 8.4. This value chain follows the delivery of content to the end-user's device, and classifies services into three categories:

1. Access-focused approach: This pertains to data services for customers who are internet-experienced. These customers tend to bypass mobile portals in order to directly access their preferred content sites. Their main requirements are wireless internet connections, out-of-office employees accessing company's network, and so on.

2. Portal-focused approach: This category concerns data services for users who are less technically savvy and have less internet experience. These customers demand simple interfaces and tend to access information via service provider's portal. Examples include receiving the weather or stock market news directly via the providers portal site.

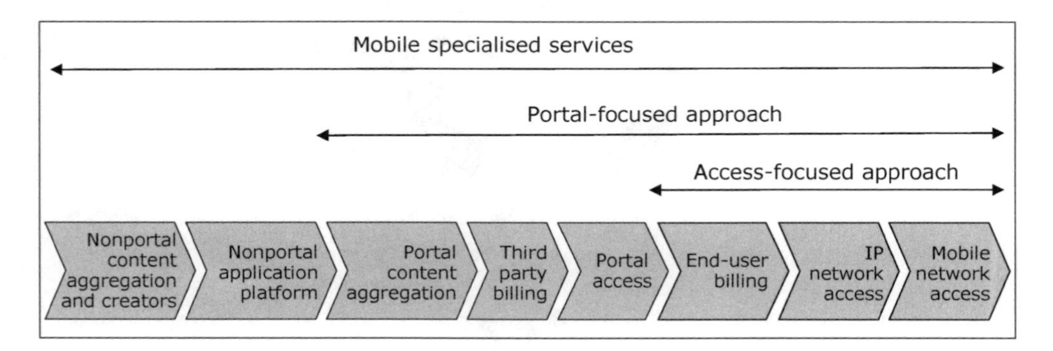

Figure 8.4 A mobile-commerce value chain. Reproduced by permission of UMTS Forum

3. Specialised service: This category pertains to services for content which has been generated by the end-users themselves. These services can include end-to-end file transfer, or messaging, or plain old telephony.

The above model is used to discuss aspects of the possible services evolution for 3G applications. It is also a good guide for broadband communications services. Following is a brief description of the different stages of this value chain:

8.2.1 Nonportal content aggregations and creators

These players generate content, which is distributed and used independently of the operator's portal. End-user generated content, such as voice (mobile telephony), short messaging service and multimedia messaging service (SMS and MMS) are already supported by various mobile functions. However, there are other content that can be generated by end-users; these could be advertising, MMS with special content, sports or security content generated by private cameras and single-cast or multi-cast over the mobile network. Ring-tones and screen savers are other popular services in this segment. Yet another major service is games, developed by many players large and small over the platforms such as Java.

Players in this segment (1) develop content using software platforms, such as Java and XHTML, or (2) provide expertise in adapting other customers' content for mobile platforms. This is a nascent business, with many small start-up companies.

8.2.2 Nonportal application platforms

These players provide application platforms over which nonportal content may be aggregated and delivered. Already, most mobile manufacturers develop or provide such software platforms. Sun Microsystems' Java is a popular platform over which games and mobile software accessories are developed. Another example is Qualcomm's BREW, which is used in many 3G phones.

8.2.3 Portal content aggregations

These players aggregate content that is delivered via an operator portal. In Japan, this is usually done by the operator who both controls the portal and aggregates the content for the portal. Application and content providers develop these for the operator, customised according to their portal platform.

8.2.4 Third party billing

Business-to-business transactions at present, are typically handled directly between the operator and content providers. The end-user pays for portal content on a fee basis or usage basis. From this revenue, the operator keeps a certain commission of $10 \sim 30\%$ and passes the balance on to the content originator.

8.2.5 Portal access

Portals are the entry point to services and applications delivered to a mobile user. An operator provides content from its partners through its portal. Portal development, control of its content, and customer management functions, can be classified under this category. The delivery of customised information and entertainment can be initiated and managed here. Most current operators carry out this function in-house.

8.2.6 End-user billing

Although at times outsourced, end-user billing is usually handled by a department or a subsidiary of an operator. As currently costs for end-users are calculated on the basis of airtime or data packet use, a measure of these usage figures are necessary. With the emergence of flat-rate services, it can be expected that these services are outsourced more often.

8.2.7 IP network access

The internet service providers (ISPs), as their name implies, provide access to the internet. Usually, fixed-line ISPs also provide a service for wireless users through contracts with the operator. For example, KDDI in Japan lists more than 30 ISPs as partners in providing internet access (see Box: Wireless ISPs). This function is billed on the basis of subscription, or as part of a flat-rate service package.

8.2.8 Mobile network access

This traditional aspect of a mobile operator's business is typically billed on the basis of usage calculations or on flat-rate terms and conditions.

8.3 Service Classifications

A framework has been developed for the UMTS Forum to forecast future services and revenue opportunities for 3G services, as illustrated in Figure 8.5.[1] While this framework is designed for the relatively lower rate 3G services, it is still applicable to the services of broadband wireless systems in general. The reason is that, generally, broadband services, fixed or wireless, have been designed as bridges connecting computers to the internet.

[1] See UMTS Forum Reports Nos. 9, September 2000.

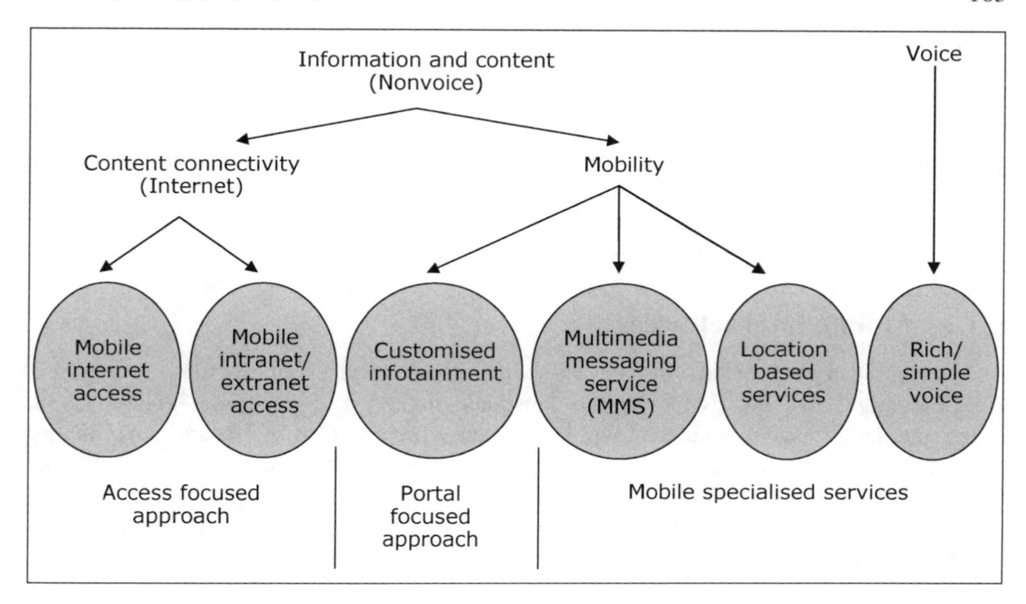

Figure 8.5 The UMTS Forum framework for 3G services. Reproduced by permission of UMTS Forum

Initially, broadband wireless is targeting end-users with portable devices, such as laptop PCs whose primary function is not communications. As illustrated in Figure 8.1, these devices are better defined as portable rather than as mobile. On the other hand, mobile communications systems, such as those of 1G, 2G, and 3G have provided voice service, initially, and then gradually evolved to offer data services. As the two market segments tend to increasingly overlap, we think the definition of services, as shown below, is applicable to both broadband wireless and mobile communication services.

Broadly, services are classified into two groups: (1) voice services and (2) all information and content-driven services, characterised as IP-based and always-on. This latter group is usually referred to as nonvoice, a designation that indicates the importance of voice traffic to an operator's business over the foreseeable future.

In total, six service categories are defined below. The first five categories are nonvoice and the last, sixth category, is voice.

1. Mobile internet access
2. Mobile intranet/extranet access
3. Customised infotainment
4. Multimedia messaging service (MMS)
5. Location-based services
6. Rich voice and simple voice

The first two service categories are access focused, in that users subscribe to these services in order to access the internet or a corporate intranet. For these services, the broadband wireless operator appears as a wireless ISP. The third category is a portal-focused service, where end-users are provided with specific content by several providers as aggregated by a mobile portal. The fourth, fifth, and sixth categories can be defined as special services which are usually user(s)-to-user(s), or business-to-user(s). In the following section, we discuss these categories in more detail, giving examples of present services, how they may be billed, and how these services may evolve in future broadband systems.

8.3.1 Mobile internet access

This service category enables mobile access to internet services as provided by a typical ISP. The target customer is mainly the consumer market; it augments or replaces fixed, wired internet access. In broadband wireless systems, transmission speeds and other quality parameters will be comparable to fixed systems. Typical services include web browsing, downloading and uploading files, email, and audio and video streaming.

Typical transmission rates for each of these services, with consideration to traffic asymmetry for uplink and downlink are shown in Table 8.1.

Traffic asymmetry

The asymmetry between the amounts of mobile internet access traffic in downlink and uplink are estimated by UMTS forum to be larger by 10 to 13 times in the downlink as compared to the uplink by the year 2020. That is, much higher capacity is needed in the downlink compared to the uplink.

8.3.2 Mobile intranet/extranet access

This service category is similar to the mobile internet access category above, with the difference that it is targeted at the business customer. This service enables the business customer to securely connect to his/her corporate network. Required QoS parameters and transmission rates are similar to those of Table 8.1.

Both the mobile intranet/extranet access and mobile internet service access can be defined as 'internet-centric'. The subscribers of these services usually augment their fixed internet access with mobile access, and expect a similar quality of service. Because they

Table 8.1 Typical transmission rates required

	Downlink	Uplink
Mobile internet access	500 kbps ~ 1 Mbps	32 kbps ~ 64 kbps
Mobile intranet/extranet access	500 kbps ~ 1 Mbps	128 kbps ~ 256 kbps
Customised infotainment	64 kbps ~ 128 kbps	8 kbps ~ 16 kbps
Multimedia messaging service	4 kbps ~ 16 kbps	4 kbps ~ 16 kbps
Location-based services	8 kbps ~ 16 kbps	1 kbps ~ 4 kbps
Rich voice	16 kbps ~ 64 kbps	8 kbps ~ 16 kbps
Simple voice	8 kbps ~ 32 kbps	8 kbps ~ 32 kbps

typically know what content they want, they can bypass content portals to access their desired web sites directly.

Present services

Several GSM and CDMA vendors offer PCMCIA and Flash memory cards, which act as bridges connecting PCs and PDAs to the internet. These services are still quite expensive. While some operators do offer flat rate for packet calls made from a mobile phone, calls made through these internet cards are charged on a per-packet basis. The total usage charges can become quite high. Currently, maximum transmission rates for several services using these bridges are summarised in Table 8.2.

Revenues

Although the operator has a direct relationship with the end-user, that relationship can easily be bypassed. For this reason, it is typically difficult, if not nearly impossible, for the operator to benefit from a share of revenue from third party billing, or from any advertising revenue. At the same time, these services can be expected to use up a significant portion of an operator's system capacity. Naturally, charging a flat rate may not be economically viable, and related revenues would necessarily have to come from usage. Alternatively, several classes of flat rate for light to heavy users may be developed. Already in Japan, several flat-rate services exist. KDDI, for example, offers multi-step flat-rate charges as illustrated in Figure 8.6. These are for packet calls made from a mobile handset, and not from PCMCIA cards. Among the technologies listed in Table 8.2, only Willcom's PHS services in Japan are offered for a flat rate of more than $50.00 per month (see Box: PHS flat-rate). As can be seen from the table, the maximum transmission rates are well below those of other technologies. Another source of revenue is based on subscription fees, as is charged similarly by ISPs.

PHS flat rate

WILLCOM

Willcom, the Japanese PHS mobile operator, offers flat-rate plans for both voice and data communications. Voice calls between Willcom users are charged a flat ~$20 a month. Flat rate for data communications are as follows:

32 kbps	$55
128 kbps	$88
256 kbps	$110

Table 8.2 Present services' maximum transmission rates

Operator	Technology	Maximum transmission rate	Charges
Willcom	PHS	128 kbps	Flat-rate
DoCoMo	WCDMA	384 kbps	Per packet basis
KDDI	CDMA2000 EVDO	2.4 Mbps	Per packet basis

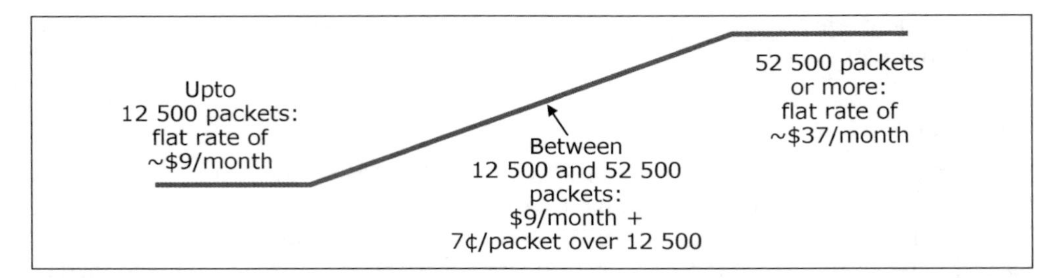

Figure 8.6 Multi-tier flat-rate service by KDDI

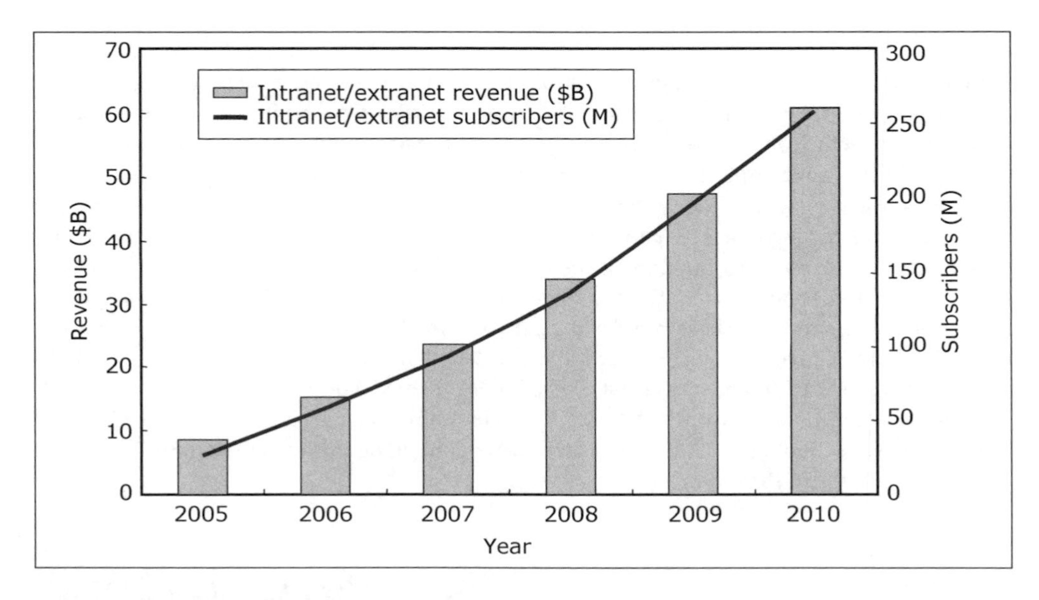

Figure 8.7 Forecast for the growth of revenues from mobile intranet/extranet access services. Reproduced by permission of UMTS Forum

Growth forecast

Wireless internet and intranet access services are expected to grow significantly over the coming years. Figure 8.7 shows the expected revenues from this service segment by the year 2010. The services are estimated to show a CAGR of 48% over the period.

Traffic asymmetry

The asymmetry between the amounts of mobile intranet/extranet access traffic in downlink and uplink are estimated by UMTS forum to be larger by four times in the downlink as compared to the uplink by the year 2020. Again, much higher capacity is needed in the downlink compared with the uplink.

8.3.3 Customised infotainment

This service category provides access to personalised information and entertainment contents to an end-user depending on the end-user's equipment. The service is geared to cater to the consumer market. Typical services in this class are m-commerce, subscription-based news, traffic reports, adult contents, and so on.

The subscribers are defined as mobility-centric, that is, their access to the internet is likely to be through specific mobile-service portal sites through their mobile devices. In fixed, wireline environments, the end-users can connect to their services using associated portals.

Present services

Most 2.5G and 3G services provide customised infotainment services. Video and audio streaming have been available for some time. In Japan, these services are provided by all major operators over their 2.5G and 3G networks. Similar services are provided over GPRS and WCDMA networks in Europe. Higher transmission rates of future networks are expected to fuel the growth of these services.

Revenues

Operator revenues are generated through resource utilisation (airtime) or flat-rate fees and commissions on portal subscription fees, advertising fees, and m-commerce transaction fees. As flat-rate services become more common, new revenue-generating models based on subscription and advertising should become more prominent.

Growth forecast

The customised infotainment services, accessible over 3G networks, are expected to grow significantly over the remaining years of this decade, with a CAGR of 38% as shown in Figure 8.8. The number of subscribers also is expected to reach 311 million, generating total revenues of $85.8 billion for operators worldwide in 2010.

Traffic asymmetry

The asymmetry between the amounts of mobile intranet/extranet access traffic in downlink and uplink are estimated by UMTS forum to be larger by seven to nine times in the downlink by the year 2020. Again, much higher capacity is needed in the downlink compared with the uplink.

8.3.4 Multimedia messaging service (MMS)

This service targets both the consumer and the business market segments. It delivers non-real-time messages to single or multiple users, with the messages consisting of text, audio, and/or video. MMS is an extension of a highly popular and profitable text-only SMS. The messages may originate from other users or from news sources, providing updates on topics such as sports, weather, and the share market.

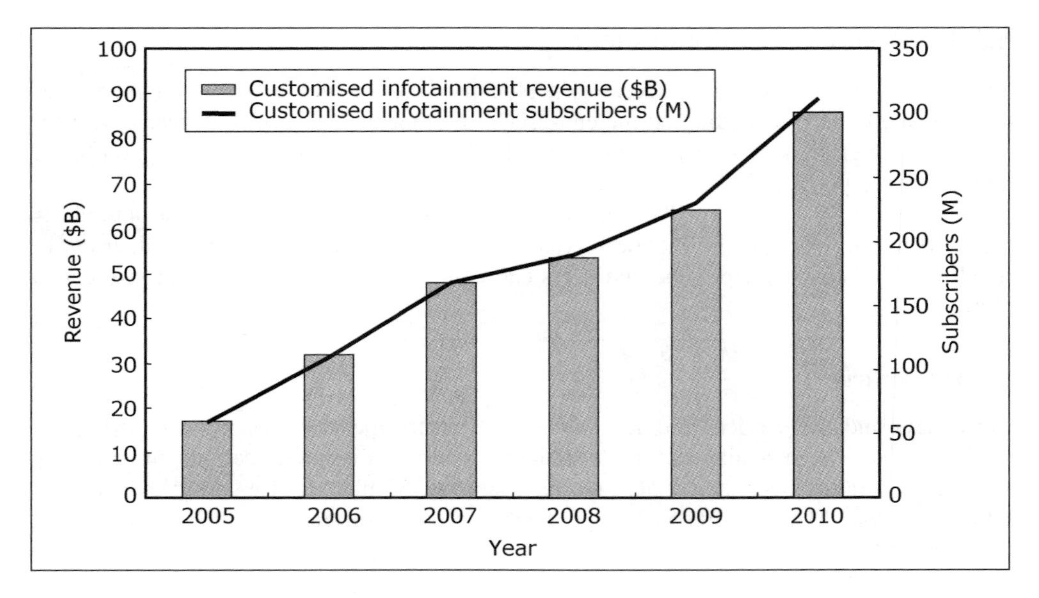

Figure 8.8 The growth of customised infotainment services revenues. Reproduced by permission of UMTS Forum

In the business segment, MMS services will provide professional users with messaging to and from mailboxes in mobile devices and laptop computers, as well as machine-to-machine communication subscriptions.

Present services

MMS has already been introduced in a limited way by operators using the 2.5G systems, with the possibility of attaching a digital photo to the message. Higher transmission rates for both 3G and broadband wireless systems should enable richer multimedia content such as audio and video clips. Various news and information SMS/MMS plans also exist.

Revenues

When MMS is user-to-user(s), the sender and/or receiver may be billed on a per-message or flat-rate basis. The operator may also receive a share of the revenue generated by the end-user subscribing to a news and information source, as a share of the subscription fee.

Growth forecast

The growth of UMTS 3G MMS services in the business segment is shown in Table 8.3. Unit of business people using MMS, and machine-to-machine services, is subscription numbers. The unit for unified messaging is the number of mailboxes. The total number in 2010 is expected to be 455 million subscriptions and units. Total MMS services over 3G networks are expected to grow significantly over the remaining years of this decade, with a CAGR of 62.2% for business MMS and 28.4% for the consumer segment, as shown in Figure 8.9.

Table 8.3 Growth of MMS services in the business segment

Segments	2005 (Million units)	2010 (Million units)
Business people using instant messaging	12	207
Machine-to-machine	105	184
Unified messaging	13.5	64

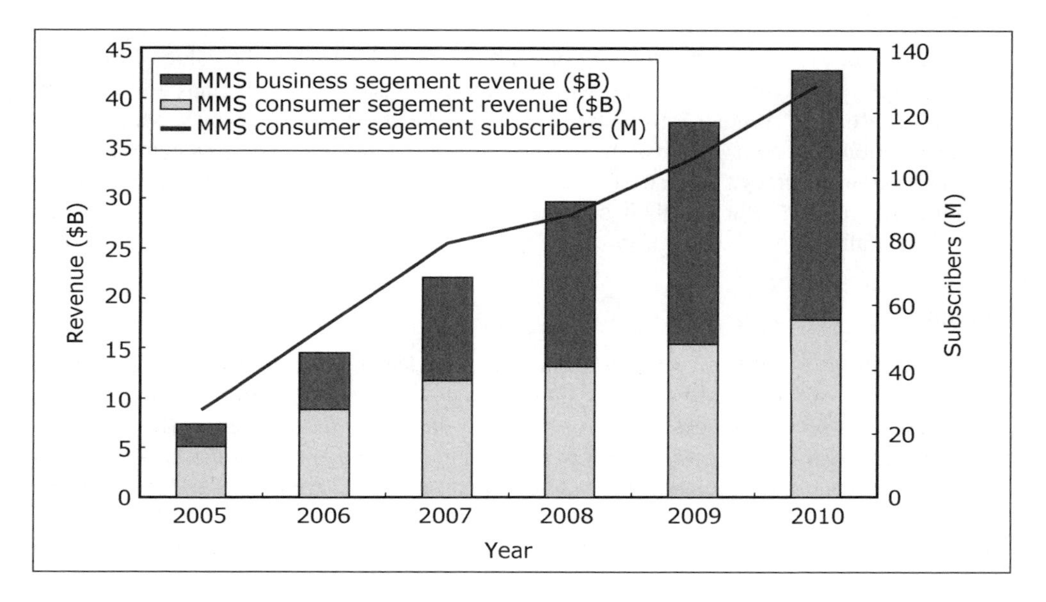

Figure 8.9 Forecast on the revenue growth of 3G multimedia messaging services. Reproduced by permission of UMTS Forum

The number of consumer subscribers is expected to reach 311 million. Total business and consumer revenues are expected to be $42.8 billion for operators worldwide in 2010.

Traffic asymmetry

The amounts of MMS traffic in downlink and uplink are estimated by UMTS forum to be almost the same by the year 2020. This is mostly because of the fact that MMS is between end-users. However, the volume of downlink traffic is slightly larger due to the fact that news may be sent in the downlink.

8.3.5 Location-based services

Location-based services exist for both the business and consumer sectors. The information on an end-user's location is valuable to both that particular user and to other businesses.

The end-user does not have to be a person. The tracking of assets, such as vehicles or goods in transit, is also an important service.

To the end user, knowing the location of a destination is useful for navigation. Other information concerning the location, for example, about the restaurants or tourist spots, is useful and can generate revenue. General information about the location on such matters as the weather and traffic patterns are further examples of services that end-users may be willing to pay for.

Present services

The accuracy of location is an important parameter which determines the value of location-based services. While navigation-type services require an accuracy of perhaps a few to tens of metres, traffic and weather information can tolerate much coarser accuracy. Many of the current location-based services use the global positioning satellite (GPS) system and have a high location accuracy. Terrestrial base station systems can triangulate the location of a mobile device, and although their accuracy is not very good, it is good enough for the delivery of traffic and weather information.

Revenues

End-users can be asked to pay for location-based services on airtime (for example, when navigating), and subscription basis. The operator may also receive additional revenue from advertisers who send messages to end-users based on their current location (for example, about restaurants that are nearby). Further revenue may be generated on the basis of transactions made; each tracking report or weather report can be charged for. The UMTS forum report estimates how service units (whether an actual human subscriber or a device tracking an asset) will grow over the period 2005–2010 for 3G services. Their forecast is shown in Table 8.4.

Growth forecast

These services over 3G networks are expected to grow significantly over the remaining years of this decade, with a CAGR of 30% as shown in Figure 8.10. The number of total subscriber/subscriber units from Table 8.4 are expected to reach 513 million, and generate total revenues of $9.9 billion for operators worldwide by 2010.

Table 8.4 Growth of location-based service units

Segments	2005 (Million units)	2010 (Million units)
Consumer navigation and guidance	22	187
Location-based advertising	39	117
Location-based transactions	4	19
Consumer third party	18	180
Business asset tracking	1	10

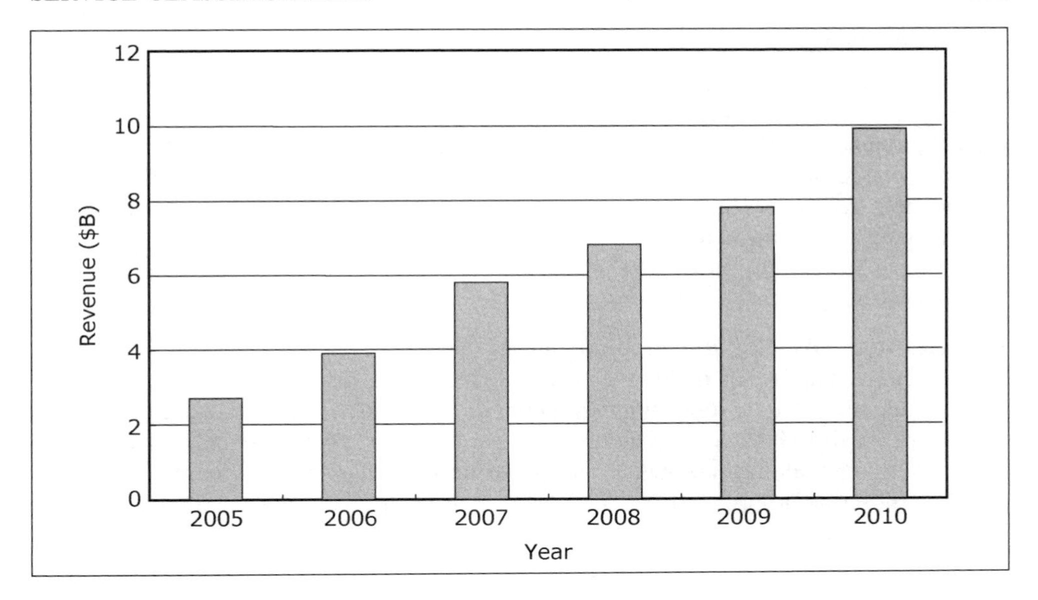

Figure 8.10 Forecast on the revenue growth of 3G location-based services. Reproduced by permission of UMTS Forum

Traffic asymmetry

The asymmetry between the amounts of traffic for location-based services in downlink and uplink are estimated by UMTS forum to be larger by seven times in the downlink by the year 2020. Again, much higher capacity will be needed in the downlink compared with the uplink.

8.3.6 Rich voice and simple voice

Voice communications is a user-to-user service, which has been a main source of revenue for mobile operators. It covers both consumer and business market sectors. As we had discussed previously, the revenue per user for simple voice services has been shrinking because of operator competition and replacement by other services such as SMS. Future 3G and broadband wireless services need to enhance the simple voice service in order to maintain present revenue levels. The terminology 'rich voice' is coined to cover possible enhancement of the old simple voice service. The following enhancements are already implemented or expected to be implemented in the near future.

Rich voice services are expected to include:

- Higher quality voice: Use of higher AMR rates to deliver higher PQoS (see Section 3.9).
- Voice over IP
- Video phones
- Push-to-talk (one-to-many) voice calls

Present services

Present services are primarily circuit-switched voice. Although new voice over IP (VoIP) services are starting to emerge, the traditional voice delivery system is expected to retain its present prominent market share. The most recent addition to voice services is Push-to-Talk. This is a one-to-many service, where one person's voice message can be delivered in real time to many receivers. Video conferencing, multimedia imaging, and photo enhanced calls are emerging services which are expected to grow.

Revenues

Revenues from voice services have traditionally been calculated on the basis of airtime. For the future, flat-rate or multi-level flat-rate charging is being considered. Charges for push-to-talk service (one-to-many) may possibly be calculated taking into consideration the special service plans and the numbers of end-receivers. Other services, such as video conferencing or multimedia imaging and photo may be charged on a flat-rate or regular monthly subscription basis, easing the operators from the per-minute billing to a more convenient flat-rate charging mode.

Growth forecast

The growth forecast for simple voice services over UMTS 3G is shown in Figure 8.11. This figure also shows the total revenue forecast for both simple and rich voice services. Rich voice subscriber numbers are a combination of several segments. Business rich voice, including web-based multimedia conferencing, is expected to be 3% of the total market share for such services (wired or wireless) in 2010 as against 0.2% in 2005. Subscribers for consumer videophone, and consumer multimedia imaging and photo services are expected to reach 12 million and 43 million respectively by 2010. Total revenues for voice services over UMTS 3G networks are expected to grow significantly over the remaining years of this decade, with a CAGR of 95.6% for rich voice and 36.3% for simple voice, as shown in Figure 8.11. The number of subscribers is expected to reach 630 million, generating total revenues of $108.6 billion for operators worldwide in 2010.

Traffic asymmetry

The amounts of simple and rich voice traffic in downlink and uplink are estimated by UMTS forum to be almost the same in the year 2020. This is again mostly owing to the fact that voice services are between end-users. In similarity with MMS services, again the downlink traffic is slightly larger because of the fact that calls from the fixed network can have a higher video transmission rate compared with the uplink streams (due to mobile handset capabilities).

8.4 Total Revenue Forecast

Table 8.5 lists in what units each service may be billed. The total revenue forecast for UMTS 3G services can be found by adding the forecasts of all the services. These are shown in Figure 8.12. A revenue compound annual growth rate (CAGR) of 41% is estimated for the period 2005–2010 in total for all services.

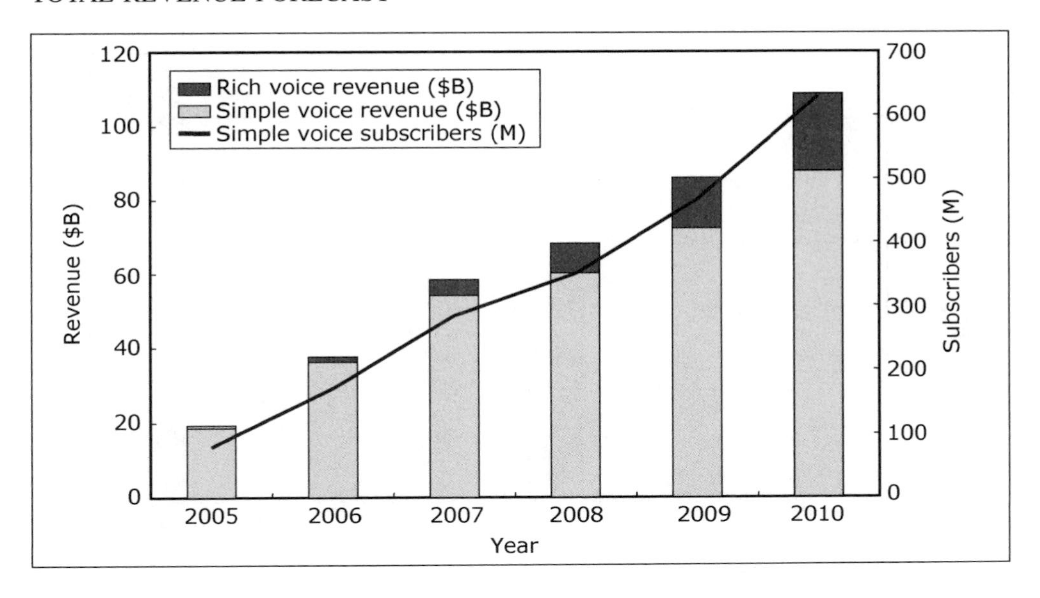

Figure 8.11 Forecast on the revenue growth of 3G simple and rich voice services. Reproduced by permission of UMTS Forum

Table 8.5 Revenue collection units. Reproduced by permission of UMTS Forum

Service	Mobile internet access	Mobile intranet/ extranet access	Customised infotainment	Multimedia messaging Service	Location-based services	Rich voice and simple voice
Airtime	√	√	√		√	√
Message				√		
Subscription	√	√	√	√	√	
Advertising			√		√	
Transaction			√		√	

Since two of the technologies that are currently being considered for broadband wireless are of UMTS 3G standard, the figures presented in this chapter are well indicative of how broadband services may grow. Current trends indicate that, at least in the short term, non-3G broadband technologies will target the wireless internet market. The forecast figures presented in this chapter are used in Chapter 9 to develop fictitious business scenarios for new operators in Japan and China.

8.4.1 Flat rate

We end this section with a review of whether it is possible to provide wireless services at flat-rate charges. The question is whether other sources of revenue, such as content subscription, advertising, and transaction fees can compensate for the loss of revenue because of flat

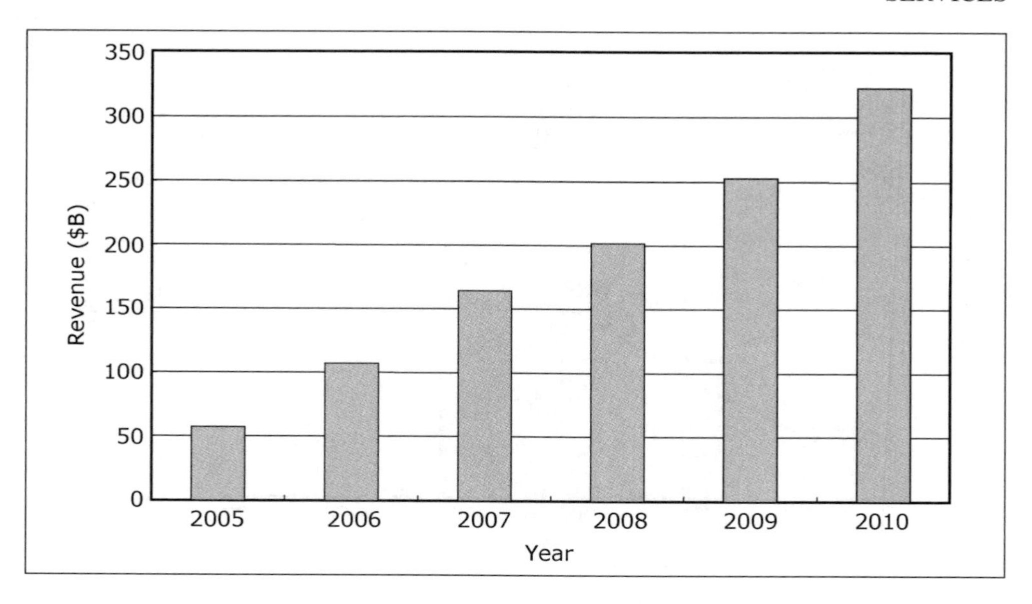

Figure 8.12 Forecast on total revenue growth of UMTS 3G services, 2005–2010. Reproduced by permission of UMTS Forum

Table 8.6 Possibility for offering flat-rate service

Service	Mobile internet access	Mobile intranet/ extranet access	Customised infotainment	Multimedia messaging service	Location-based services	Rich voice and simple voice
Flat-rate possibility	√	√	√√√	√√√	√√√	√√

rate. As shown in Table 8.5 many of the revenue forecasts for the service categories of this chapter are calculated on the basis of airtime. However, with relatively smaller traffic volumes, it should be possible to charge a flat rate for most of these services. It is reasonable to assume that those services that must go through an operator's portal can become flat-rate. The reason is that the operator should then be able to devise different models for generating revenue. Only Mobile internet, intranet and extranet access services can bypass the operator portal; furthermore, they require relatively larger bandwidths. These services can be charged on a multi-tier flat-rate basis as, for example, KDDI's Double Teigaku as shown in Figure 8.6.

Voice service is another category that has traditionally been charged on the basis of airtime. Here too, users may be offered superior service packages at fixed subscription rates, which could include higher quality rich voice.

Table 8.6 summarises this author's belief on the extents to which each of the services can be charged at flat rate. It states that all services can be charged on a flat rate basis.

However, mobile internet and intranet access, as well as voice flat rate charging may have to be done on a several-tier basis.

Summary

In this chapter, we discussed the services and possible revenues that could be expected from future broadband wireless communications services. We further discussed about possible revenue flow scenarios and wireless communications business value chains. Using a model and figures developed by UMTS forum, we presented forecasts about communications services over the 2005–2010 period. We will use these figures to develop business models for new wireless operators who will be using the technologies discussed in this book.

Further Reading

- On Mobile commerce:
  - Paavilainen, J., *Mobile Business Strategies*, IT Press, 2001.
  - Sadeh, N. M., *M-Commerce: Technologies, Services, and Business Models*, John Wiley & Sons, 2002.
- On Subscriber estimates for 3G systems UMTS Forum Reports:
  - UMTS Forum Report No. 9, *The UMTS Third Generation Market Structuring the Service Revenues Opportunities*, 2000.
  - UMTS Forum Report No. 13, *The UMTS Third Generation Market. Phase II: Structuring the Service Revenue Opportunities*, 2001.
  - UMTS Forum Report No. 37, *Magic Mobile Future 2010–2020*, 2005.

9

Scenarios

Following the introduction to new technologies being utilised in 3G systems and beyond, in Chapters 2, 3, and 4, we have given a summary of the costs associated with building and operating a wireless mobile communications system in Chapters 5, 6, and 7. With a list of possible services provided over these systems given in Chapter 8, we now will develop several business scenarios for fictitious operators who may utilise these new technologies to provide wireless services.

Thus, this final chapter is intended to demonstrate the behaviour of factors that go into the making of such a wireless business model. These factors, such as the initial investment required, cost of capital, target market and its size, growth factors, expected services, and so on, are introduced and calculated. The technologies introduced and discussed in this book have characteristics that make them suitable for one type of service or another. Although it may be technically possible to provide any service using any of these technologies, we believe there are certain technologies more suited to a particular service type. This belief comes from the fact that some technologies were initially designed for a particular market need and, therefore, are better suited to serve that need.

The following discussion concentrates on four technologies. Of these four, our focus is mostly on the first three as we believe that the fourth technology is neither well equipped nor intended for wide-area coverage.

1. WCDMA
2. TD-CDMA
3. WiMAX
4. WLAN.

We have chosen two target markets, Japan and China, for developing business scenarios. The first target is Japan, mainly because it is arguably the world's most advanced market in terms of mobile communications services, particularly data services. Japan is also chosen because the author lives in Japan and is most familiar with the Japanese market. The reasons

Broadband Wireless Communications Business: An Introduction to the Costs and Benefits of New Technologies Riaz Esmailzadeh

for the second choice are that China is the largest telecom market, has an underdeveloped fixed telecom infrastructure, and is growing fast. In discussing these two markets, we have tried to use figures that are up to date, and forecasts that are realistic. It should be kept in mind, however, that forecast accuracy is not a major criterion in a textbook in the same way that it is perhaps required in the report of a financial analyst. We have tried to use realistic assumptions, although the assumptions are sometimes simplistic. The main intention is to familiarise the reader with methodologies used for developing business models. Prospective operators are expected to develop more complex market strategies. Thus, the author hopes that the following discussion gives an indication of the factors considered in an operator's business plan. We also intend to demonstrate the aim of this book as described in the preface: to show how the choice of a technology leads to different business models and service offerings. The reverse is also true, of course; if an operator intends to provide a particular service to a specific target market, then the choice of technology is very important.

This chapter is organised as follows. First, we have listed the assumptions for the technologies used in different scenarios. Then we have listed the market size for a set of possible services in each country based on the service and growth models discussed in Chapter 8. Next, the assumptions and costs of building a network infrastructure for each technology and, based on that, the required capital expenditures are listed. Assuming certain operational expenditures, and extrapolating from these assumptions and figures, we then evaluate the potential revenues based on a market share scenario.

9.1 Technologies

As stated earlier, four technologies are being considered overall: WCDMA, TD-CDMA, WiMAX, and WLAN. The last technology is examined only in brief as it is not well suited, nor intended, for wide area coverage (as its name, wireless *local* area network, also suggests). We assume an equivalent total bandwidth, 10 MHz, for the first three technologies. For FDD systems, this is divided into 2×5 MHz for uplink and downlink transmissions, whereas for the TD-CDMA systems, it is a full 10-MHz block dynamically used for duplex transmissions. We further assume that the cost of spectrum is identical, with the exception of WLAN systems, which use unlicensed bands. The WLAN operator is assumed to have a 20 MHz bandwidth

As the spectrum resource is finite, the maximum number of users that can be connected to a network are limited. One needs to take these limits into consideration when calculating the maximum subscriber numbers and revenues. These limits, as well as other system-specific parameters, are listed in the following text.

9.1.1 WCDMA solution

Within the present release of the WCDMA standard, each carrier bandwidth is set to 5 MHz. With a 10-MHz bandwidth available, one downlink carrier and one uplink carrier can be used. While much has been made of maximum throughput for WCDMA systems, it is the average throughput figure that we take as indicative of the useful bit rate for end-users. Systems assumptions for the different simulations are listed in Table 9.1.

Table 9.1 WCDMA system parameters (assumptions)

Bandwidth	10 MHz (5 uplink, 5 downlink)
Adaptive array antennas	No
Joint detection	No
HSDPA	Yes
All-IP backbone	Yes
Coverage area	Nation-wide

9.1.2 TD-CDMA solution

For the TD-CDMA system, we use a 10-MHz bandwidth. Again, we take the average throughput figure as indicative of the useful bit rate for end-users. Systems assumptions for the different simulations are listed in Table 9.2.

9.1.3 WiMAX solution

The WiMAX system bandwidth is also assumed to be 10 MHz. Although WiMAX allows for both FDD and TDD operations, we only consider the FDD here. Also, we use the average throughput figure as indicative of the useful bit rate for end-users. Systems assumptions for the different simulations are listed in Table 9.3.

9.1.4 WLAN solution

The WLAN system has several typical bandwidths. Here, we shall assume a 20-MHz bandwidth, as used in the 802.11a standard. We note that the allocated, unlicensed bandwidth is twice that of the other systems mentioned earlier. The CSMA character of the WLAN

Table 9.2 TD-CDMA system parameters (assumptions)

Bandwidth	10 MHz (uplink and downlink)
Adaptive array antennas	No
Joint detection	Yes
HSDPA	Yes
All-IP backbone	Yes
Coverage area	Dense urban, and then nation-wide

Table 9.3 WiMAX system parameters (assumptions)

Bandwidth	20 MHz (5 uplink, 5 downlink)
Adaptive array antennas	No
Joint detection	No
Multi-level modulation and coding	Yes
All-IP backbone	Yes
Coverage area	Rural, urban and then nation-wide

Table 9.4 WLAN system parameters (assumptions)

Bandwidth	20 MHz (uplink and downlink)
Adaptive array antennas	No
Joint detection	No
Multi-level modulation and coding	Yes
All-IP backbone	Yes
Coverage area	Hot spot

systems leads to uncertainties about their average throughput figures. We use the figures as stated in Table 5.13. These, and other system assumptions are listed in Table 9.4.

9.2 Market Size

For some services, the target market may be a certain age group of the population. Other services may classify the population into working segments, such as businessmen/women or general consumers. While the data unit for some of these services is the individual, other services are measured in terms of households as the data unit. For example, mobile phone market penetration is calculated as the number of phones divided by the population. The market may be subdivided by age group and the corresponding market penetration calculated accordingly. In contrast, it is more meaningful to present the fixed DSL market penetration in units of homes, or households.

A wireless operator may design a service aimed at a particular market segment. The maximum size of these markets can easily be found in relevant statistics, including numbers of households, population segments, etc. The market size may further be narrowed by considering factors such as place of residence (urban or rural) and income levels. The operator may then decide how to serve the target market, leading to critical decisions on the required system coverage, initial capital expenditure, and so forth.

The same targeting style can be used in the case of machine-to-machine communications, which, as described in Chapter 8, is an important growing market segment. Here, again, the size of the market can be found from statistics, from which the targeted market segment is estimated.

This chapter concentrates on only two markets as stated earlier, Japan and China. For both these markets, we use population projections and subscriber growth figures for mobile communications services. For DSL and wireless DSL, we consider the number of households, and make some assumptions regarding possible growth rates. Our modelling time frame is set for the years 2005–2010 and predicts the sizes of these two markets in this period. On the basis of these market size figures, we compare the cost and revenue for the four technologies being discussed.

9.2.1 Japanese market

The Japanese population has almost reached a plateau, with population growth in recent years being less than 1%. Figure 9.1 shows the expected population size over the years

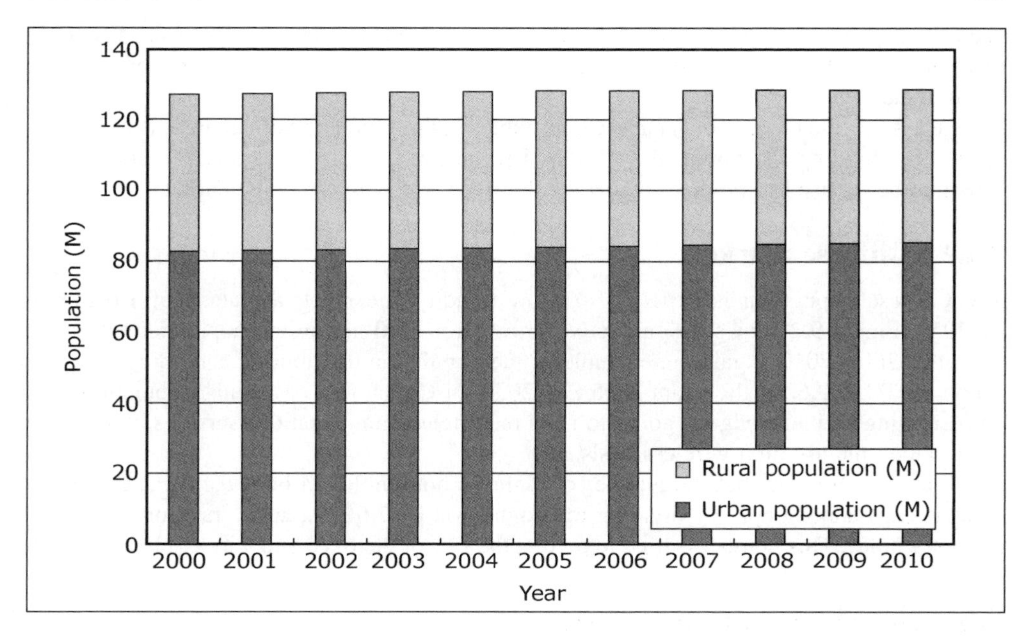

Figure 9.1 Japanese population forecast, 2005–2010

Table 9.5 Japanese population age distribution in 2005 and 2050 (Source: United Nations Population Division)

Age	Percentage of the population 2005 (%)	2050 (%)
0–10	9.3	8.9
10–20	9.8	9.1
20–30	12.6	9.1
30–40	14.6	9.7
40–50	12.4	10.6
50–60	14.9	10.9
60–70	12.4	12.2
70–	14.0	29.5

2000–2010. Rural and urban populations are expected to remain in similar proportion, with rural population declining slightly and urban population increasing slightly. The source of these figures is the United Nations Population Division.

In building market size models, population distribution according to age is important. As an example, the age distribution of the Japanese population for the years 2005 and 2050 is shown in Table 9.5. An ageing population, with increasing IT knowledge and IT needs, is expected to result in growth for the telecommunications market. This need is going to be filled partly by fixed, optical fibre connections to the home, and partly by wireless technologies.

We assume the number of Japanese households in both rural and urban areas is calculated simply by dividing the population by 2.6 (i.e., an average of 2.6 people per household). This number is used to predict the size of the fixed and wireless DSL market.

9.2.2 Chinese market

The Chinese population is expected to grow steadily and reach a plateau of 1.6 billion by 2050. Figure 9.2 shows the respective growth for rural and urban populations over the period of 2000–2010. Again, an example of the population distribution according to age is shown in Table 9.6 for the years 2005 and 2050. In China, too, an ageing population with an increasing IT knowledge is going to need more telecommunications services, which will be provided mainly on a wireless basis.

Again, we assume that the number of Chinese households in both the rural and urban areas is calculated simply by dividing the population by 2.6 (i.e., an average of 2.6 people per household). This number is used to predict the size of the fixed and wireless DSL market.

9.3 Services and Revenues

We have defined a number of services for future wireless communications, and their expected worldwide subscriber numbers and revenues are given in Chapter 8. In this section,

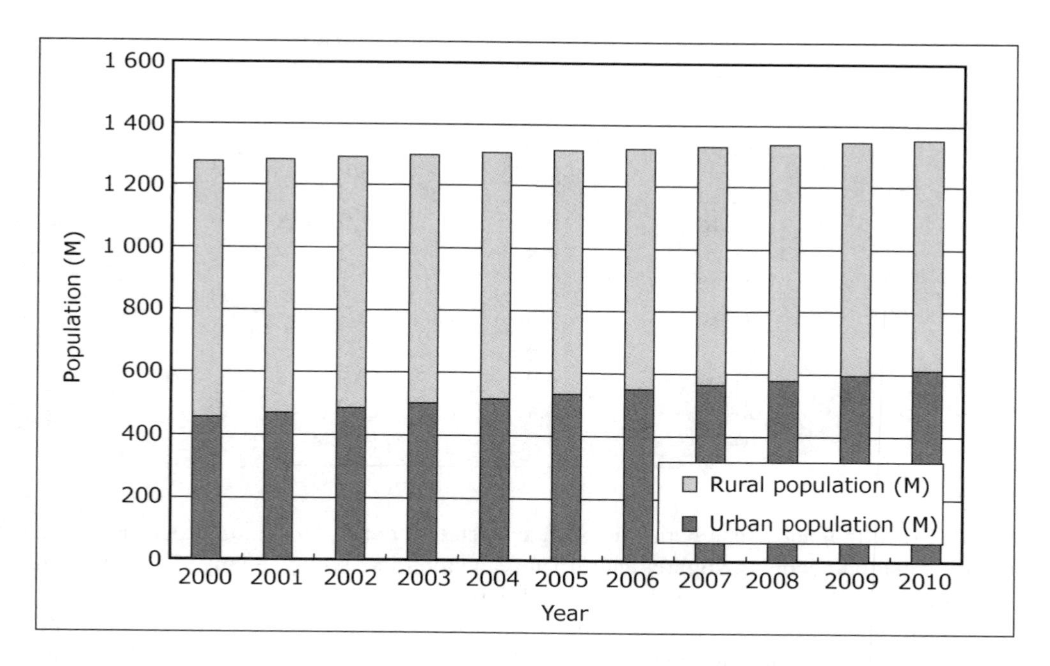

Figure 9.2 Chinese population forecast, 2005–2010

Table 9.6 Chinese population age distribution in 2005 and 2050 (source: United Nations Population Division)

Age	Percentage of the population 2005 (%)	2050 (%)
0–10	13.6	10.4
10–20	16.7	10.6
20–30	15.0	11.1
30–40	18.6	12.4
40–50	14.2	11.5
50–60	10.9	13.0
60–70	6.2	13.3
70–	4.8	17.7

we consider the following classification of these services:

1. Simple voice
2. Rich voice
3. Wireless DSL, urban
4. Wireless DSL, rural
5. Customised infotainment
6. Mobile intranet/extranet
7. Multimedia messaging service – consumer
8. Multimedia messaging service – machine-to-machine
9. Location-based services.

We use this classification as an assumption of the services that may be offered in Japan and China during the period 2005–2010. We also present the revenue figures for these services, based on the present revenue figures and some assumptions. The subscriber numbers lead to traffic figures, which, in combination with revenue estimates, can be used to develop business scenarios for our new operators. The services considered their estimated subscriber numbers, and their revenues are as discussed in Sections 9.3.1 through 9.3.9.

9.3.1 Simple voice

Figure 9.3 shows our assumptions on the number of traditional voice service subscribers in Japan and China, and the expected revenues from this service during 2005–2010. Although more than 3 $\sim$ 5 times the number of subscribers exist in China, the revenue difference is assumed to be only about twice that of Japan due to the large difference in relative ARPU figures. We assume that subscriber numbers in Japan will grow at 3% per year, and that

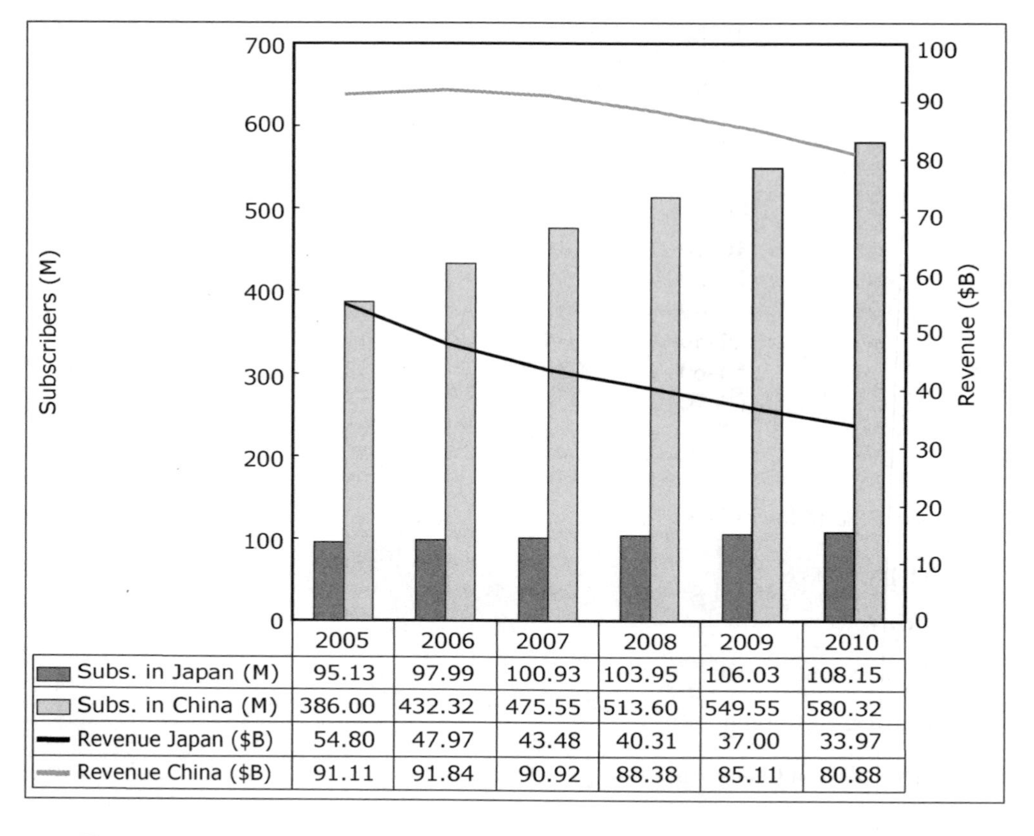

	2005	2006	2007	2008	2009	2010
Subs. in Japan (M)	95.13	97.99	100.93	103.95	106.03	108.15
Subs. in China (M)	386.00	432.32	475.55	513.60	549.55	580.32
Revenue Japan ($B)	54.80	47.97	43.48	40.31	37.00	33.97
Revenue China ($B)	91.11	91.84	90.92	88.38	85.11	80.88

Figure 9.3 Simple voice service subscriber and revenue predictions, 2005–2010

revenues will decrease by an average of 10% per year over the period. By contrast, in China, the number of subscribers will grow by an average of more than 8%, while total revenues slightly decrease.

9.3.2 Rich voice

As defined in Chapter 8, rich voice services provide higher quality voice communications enriched by features such as video, still pictures, conferencing, and one-to-many connectivity. Figure 9.4 shows our assumptions on the number of rich voice service subscribers in Japan and China, and the expected revenues from this service during 2005–2010. We have assumed that by 2010, 10% of all simple voice subscribers will also subscribe to these enriched services in both countries. Further, assuming that the CAGR for these subscriber numbers is 60% for the period, we have predicted the number of subscribers. The average revenue per subscriber would be $700 in 2005, gradually decreasing by 12% per year to $392.45 in 2010 in both countries.

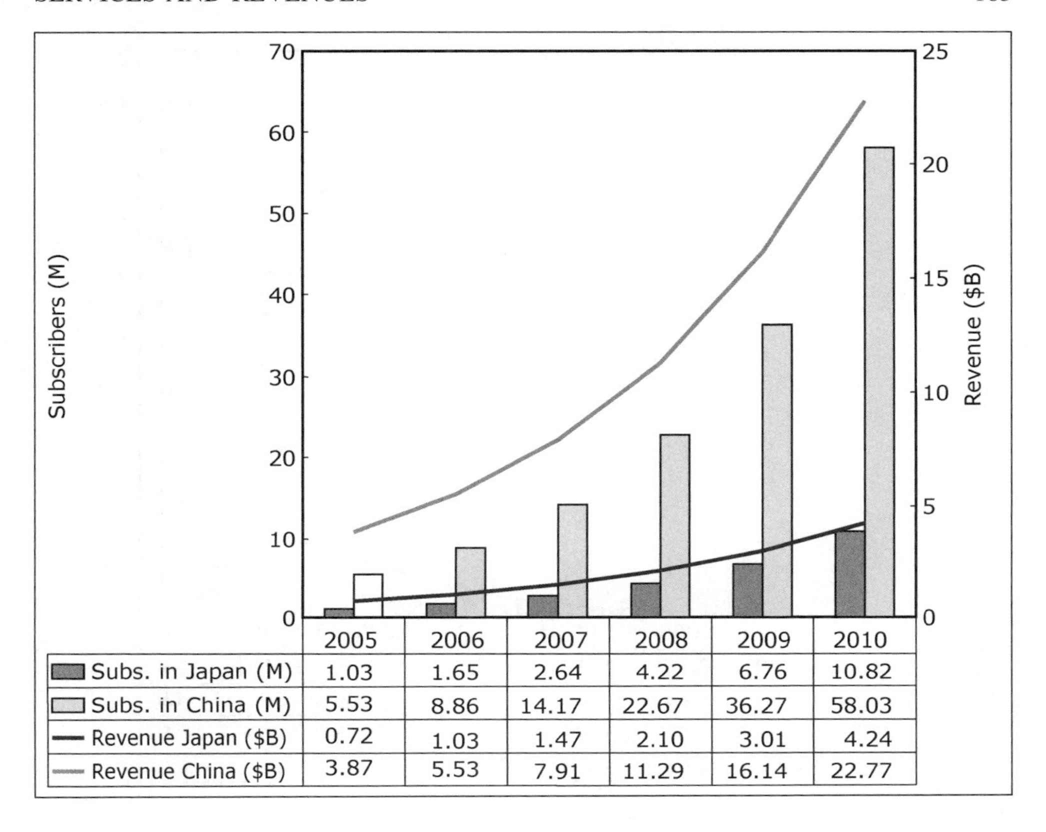

	2005	2006	2007	2008	2009	2010
Subs. in Japan (M)	1.03	1.65	2.64	4.22	6.76	10.82
Subs. in China (M)	5.53	8.86	14.17	22.67	36.27	58.03
Revenue Japan ($B)	0.72	1.03	1.47	2.10	3.01	4.24
Revenue China ($B)	3.87	5.53	7.91	11.29	16.14	22.77

Figure 9.4 Rich voice service subscriber and revenue predictions, 2005–2010

9.3.3 Wireless DSL, urban

Figure 9.5 shows our assumptions on the number of wireless DSL access subscribers in Japan and China in urban areas, and the expected revenues from this service during 2005–2010. The subscriber numbers for this service are assumed to grow at a CAGR of 139 and 274% over the period in Japan and China respectively. On the basis of these growth figures, and on a slightly decreasing revenue per user, it is assumed that the revenue CAGR would be about 120 and 249% for Japan and China respectively.

9.3.4 Wireless DSL, rural

Figure 9.6 shows our assumptions on the number of wireless DSL access subscribers for rural areas in Japan and China. The figure again shows the expected revenues from this service during 2005–2010. We assume that the subscriber numbers will grow at a CAGR of 138 and 303% for Japan and China respectively over the period. On the basis of these growth figures, and with similar revenues per user as in the urban areas, the revenue CAGRs

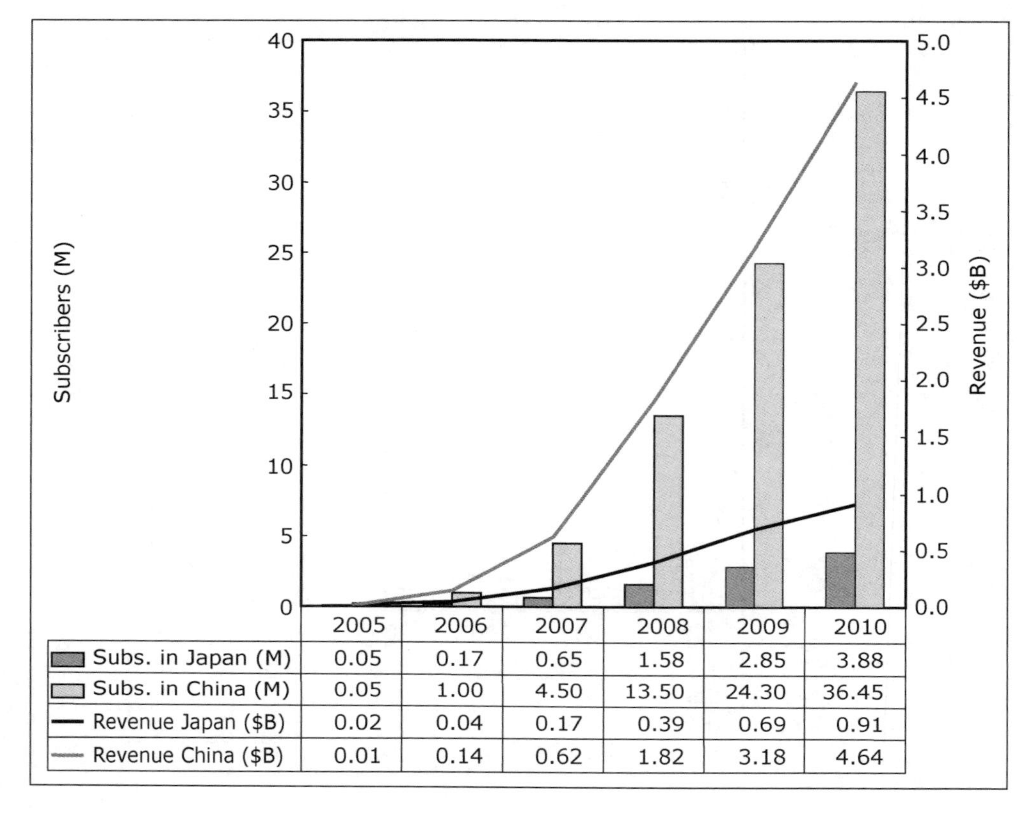

	2005	2006	2007	2008	2009	2010
Subs. in Japan (M)	0.05	0.17	0.65	1.58	2.85	3.88
Subs. in China (M)	0.05	1.00	4.50	13.50	24.30	36.45
Revenue Japan ($B)	0.02	0.04	0.17	0.39	0.69	0.91
Revenue China ($B)	0.01	0.14	0.62	1.82	3.18	4.64

Figure 9.5 Urban wireless DSL access service subscribers and revenue predictions, 2005–2010

are calculated to be 120 and 276% for Japan and China respectively. We combine urban and rural wireless DSL subscriber figures in our simulations below.

9.3.5 Customised infotainment

Figure 9.7 shows our assumptions on the number of customised infotainment subscribers in Japan and China, and the revenues from this service during 2005–2010. We have assumed that by 2010, 35 and 22% of mobile (simple voice) subscribers in Japan and China respectively will subscribe to these specialised services. Assuming a CAGR of 39.7%, and borrowing revenue predictions from the UMTS report for the Asia–Pacific region, we have calculated the number of subscribers and respective revenues for the period as shown in Figure 9.7.

9.3.6 Mobile intranet/extranet

Figure 9.8 shows our assumptions on the number of mobile intranet/extranet subscribers in Japan and China, and the revenues anticipated from this service during 2005–2010. We have

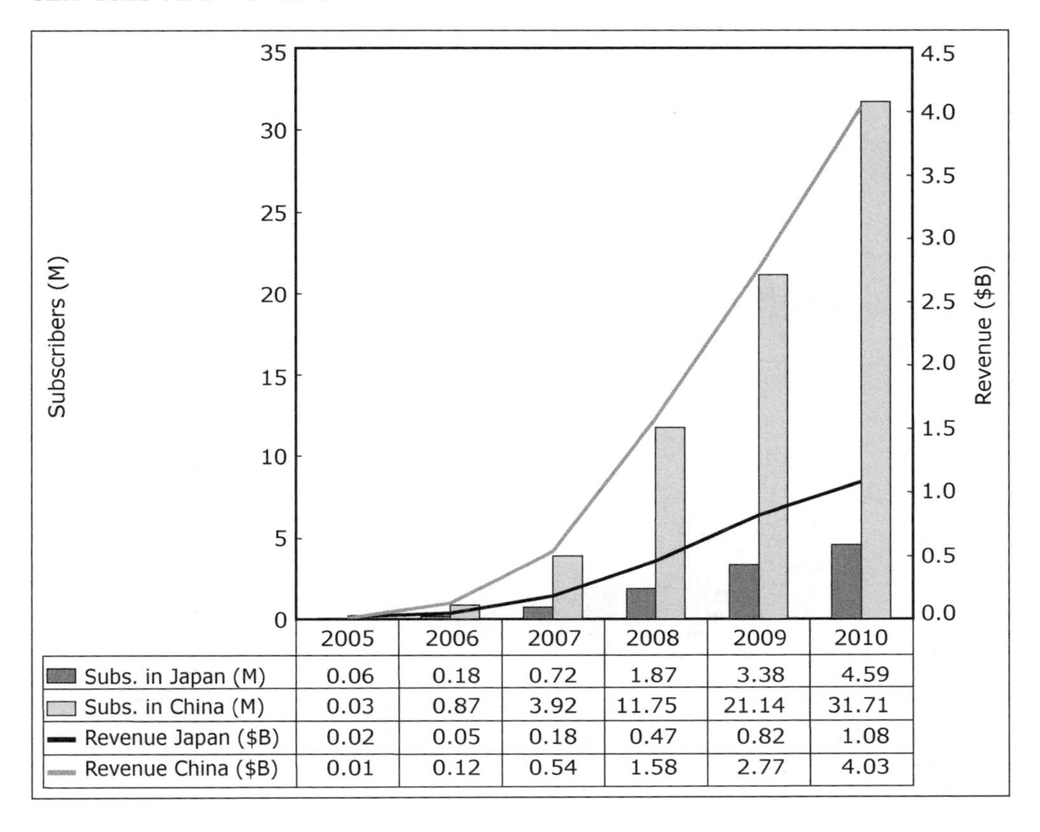

	2005	2006	2007	2008	2009	2010
Subs. in Japan (M)	0.06	0.18	0.72	1.87	3.38	4.59
Subs. in China (M)	0.03	0.87	3.92	11.75	21.14	31.71
Revenue Japan ($B)	0.02	0.05	0.18	0.47	0.82	1.08
Revenue China ($B)	0.01	0.12	0.54	1.58	2.77	4.03

Figure 9.6 Rural wireless DSL access service subscribers and revenue predictions, 2005–2010

assumed that by 2010, 20 and 10% of the total mobile (simple voice) subscribers in Japan and China respectively will use these services. Assuming a CAGR of 58.6%, and relying on revenue predictions from the UMTS report for the Asia–Pacific region, we have calculated the number of subscribers and respective revenues for the period as shown in Figure 9.8.

9.3.7 Multimedia messaging service – consumer

Figure 9.9 shows our assumptions on the number of multimedia messaging service subscribers in the consumer segment for Japan and China, and the respective revenues from this service during 2005–2010. This time we have started from 2005, and assumed that 57 and 20% of all simple voice subscribers in Japan and China respectively subscribe to this service. Assuming a 10% CAGR, we have predicted the number of subscribers for the period. Average revenue per user for this service has followed the predictions of the UMTS report number 13.

9.3.8 Multimedia messaging service – machine-to-machine

Figure 9.10 shows our assumptions on the number of devices that use multimedia messaging service to communicate with each other. We have assumed that in 2005 the number of these

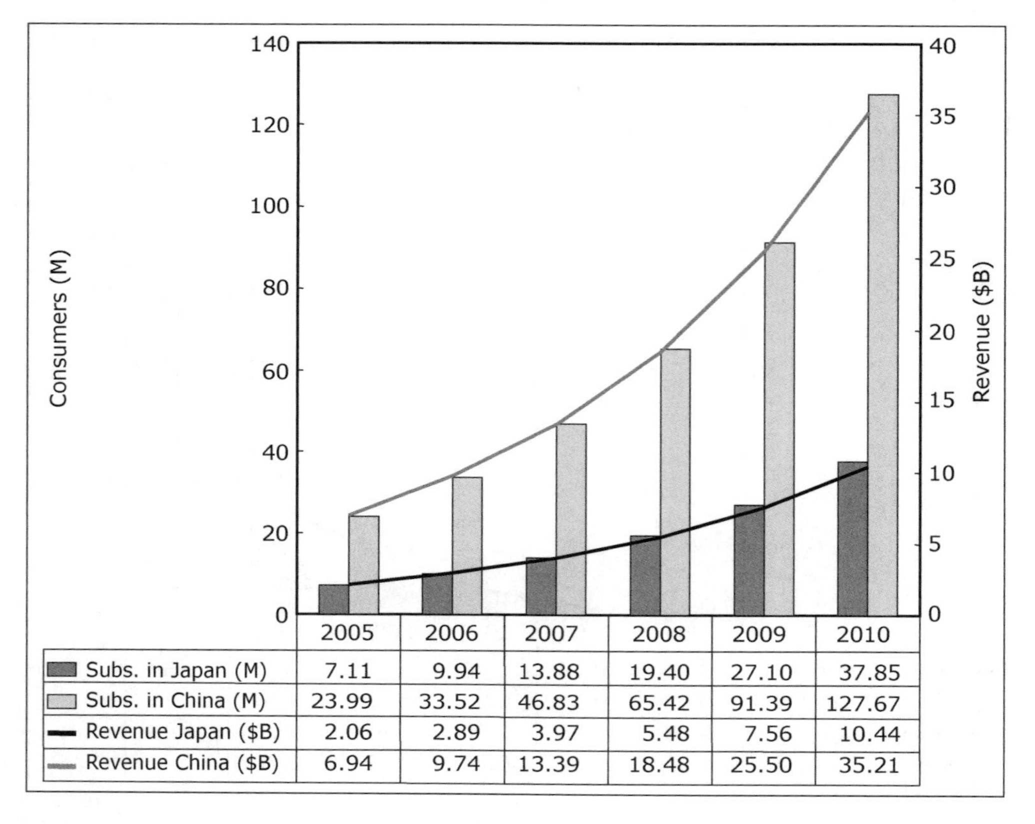

	2005	2006	2007	2008	2009	2010
Subs. in Japan (M)	7.11	9.94	13.88	19.40	27.10	37.85
Subs. in China (M)	23.99	33.52	46.83	65.42	91.39	127.67
Revenue Japan ($B)	2.06	2.89	3.97	5.48	7.56	10.44
Revenue China ($B)	6.94	9.74	13.39	18.48	25.50	35.21

Figure 9.7 Customised infotainment service subscribers and revenue predictions, 2005–2010

devices are 1 million and 1.5 million respectively in Japan and China and we have further assumed that by 2010 there will be 20 million and 60 million such devices respectively. From these figures, we have calculated CAGR figures of 82 and 109% for the period. Revenues are assumed to be $360 per device per year in 2005, gradually declining to $240 per device per year in 2010.

9.3.9 Location-based services

As defined in Chapter 8, location-based services include both business and consumer segments. Information on the location of an end-user (a person or a device) is valuable to the user and to businesses (advertiser, network, tracking companies, etc). Figure 9.11 shows our assumptions on the number of end-users, both for individuals and devices, which will be using location-based services during 2005–2010. We have assumed that these services will be used by subscribers totalling to about 25 and 15% of the total mobile subscriber populations in Japan and China respectively. Borrowing the CAGR figure of 53.4%, and revenue predictions per subscriber from the UMTS report for the Asia–Pacific region, we

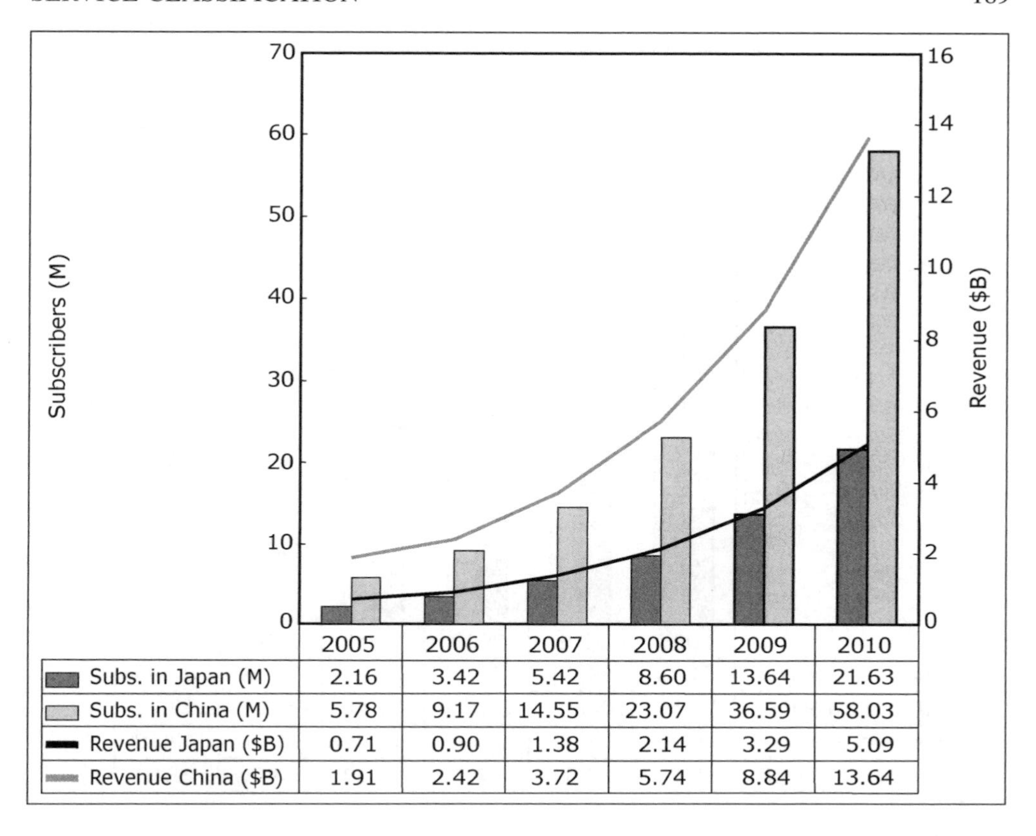

	2005	2006	2007	2008	2009	2010
Subs. in Japan (M)	2.16	3.42	5.42	8.60	13.64	21.63
Subs. in China (M)	5.78	9.17	14.55	23.07	36.59	58.03
Revenue Japan ($B)	0.71	0.90	1.38	2.14	3.29	5.09
Revenue China ($B)	1.91	2.42	3.72	5.74	8.84	13.64

Figure 9.8 Mobile intranet and extranet service subscribers and revenue predictions, 2005–2010

have calculated the number of subscribers and respective revenues for the period as shown in Figure 9.11.

9.4 Service Classification

For the purpose of market classification, and the segments that our operators will target, we classify the earlier-mentioned services into three groups:

1. Voice services: simple voice and rich voice
2. Multimedia service: customised infotainment, multimedia messaging services (consumer and machine-to-machine), and location-based services
3. Data services: wireless DSL, urban and rural, and mobile intranet/extranet.

On the basis of these three groups, our operators are presumed to enter the market focusing on a certain service group and then proceeding, step by step, to provide a broader range of services.

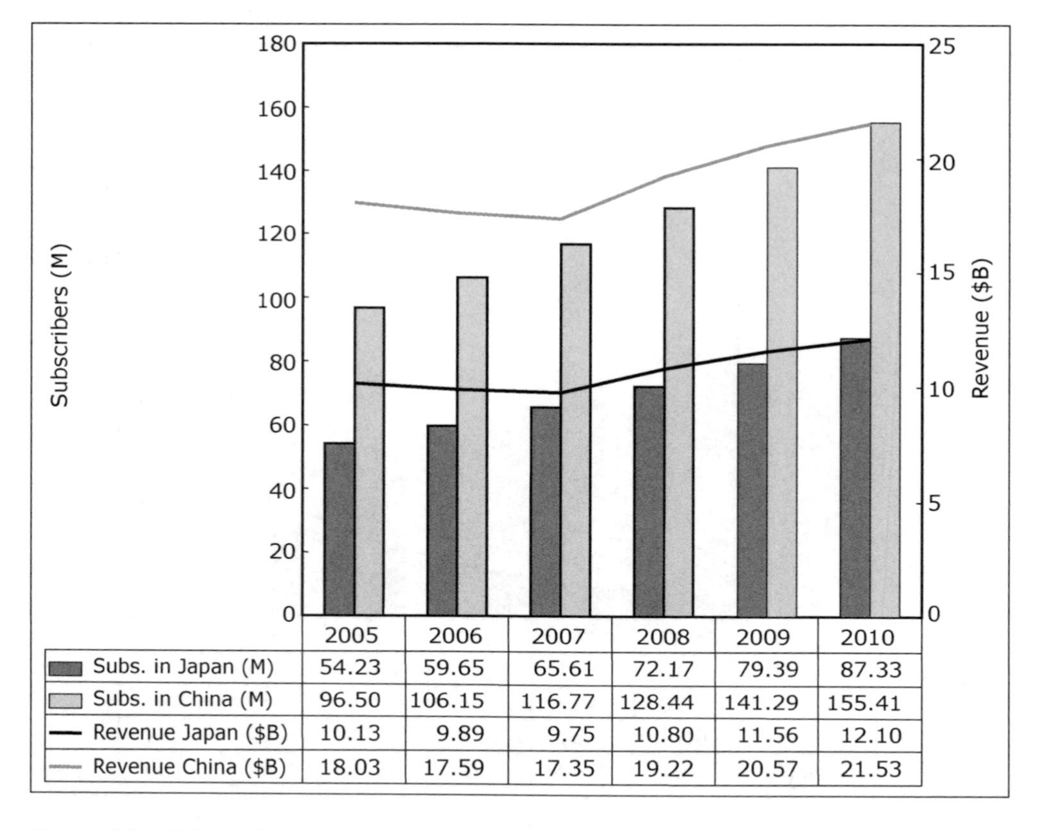

	2005	2006	2007	2008	2009	2010
Subs. in Japan (M)	54.23	59.65	65.61	72.17	79.39	87.33
Subs. in China (M)	96.50	106.15	116.77	128.44	141.29	155.41
Revenue Japan ($B)	10.13	9.89	9.75	10.80	11.56	12.10
Revenue China ($B)	18.03	17.59	17.35	19.22	20.57	21.53

Figure 9.9 Multimedia messaging service (consumer) subscribers and revenue predictions, 2005–2010

Our first fictional operator initially targets the voice service group, and then tries to extend the offering range to include multimedia and data services. This strategy is assumed for the operator using the WCDMA technology.

A second operator starts service by focusing on multimedia services and then expands into both the voice and data domains. The operator is assumed to be using the TD-CDMA technology.

A third operator competes in the data services group. Here, the operator provides a wireless bridge to the internet using any of the technologies listed earlier. The service may be fixed point-to-point, or portable. The technology is assumed to evolve to provide mobility and therefore to include voice and multimedia services. In our following simulation, this third operator uses WiMAX technology.

The fourth operator (WLAN) is also assumed to compete for services in this group, though it is not assumed to aim for the broadening of its service offerings.

These service entry points are illustrated in Figure 9.12. Our assumptions are to a great degree based on the present capabilities of these technologies as well as the range of end-user devices available as of September 2005, when these words are being written. However,

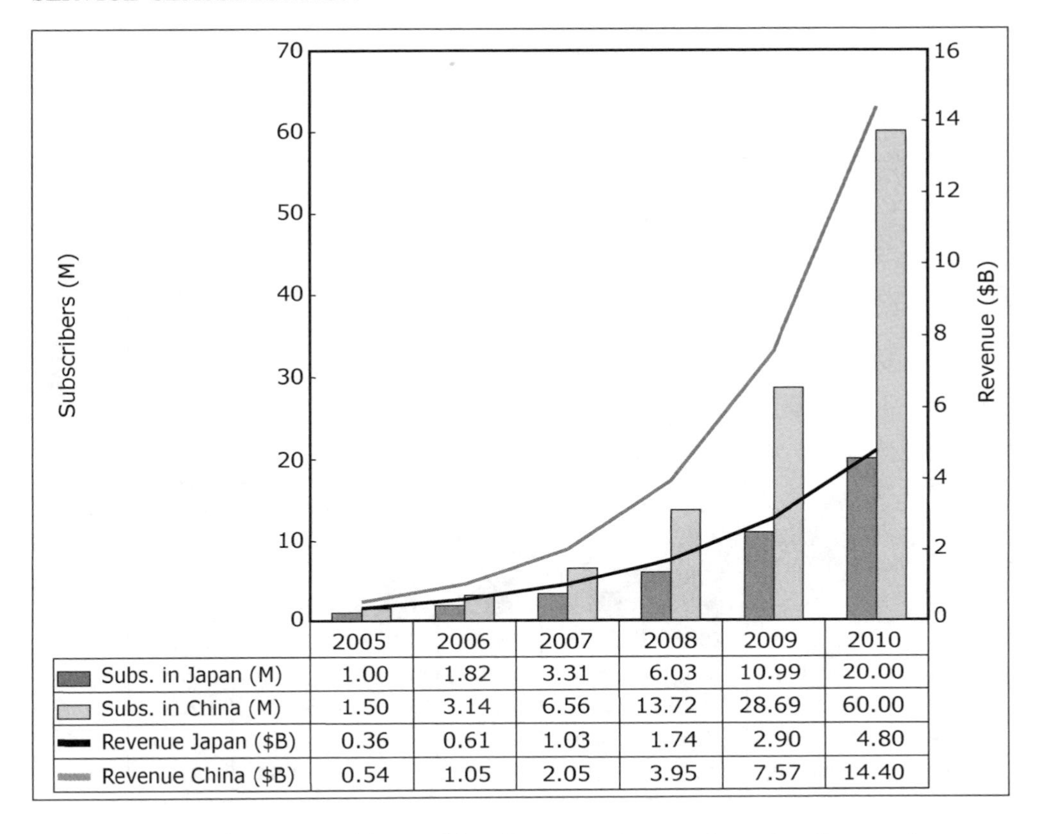

	2005	2006	2007	2008	2009	2010
Subs. in Japan (M)	1.00	1.82	3.31	6.03	10.99	20.00
Subs. in China (M)	1.50	3.14	6.56	13.72	28.69	60.00
Revenue Japan ($B)	0.36	0.61	1.03	1.74	2.90	4.80
Revenue China ($B)	0.54	1.05	2.05	3.95	7.57	14.40

Figure 9.10 Machine-to-machine multimedia messaging service subscribers and revenue predictions, 2005–2010

these are assumptions, and these and other strategic entries are possible for the selected technologies.

9.4.1 Traffic per service

A number of models exist for voice and data traffic as discussed in Section 2.8.2. Our assumptions on transmission rates for downlink and uplink traffic, as well as average holding times during the peak hour, are shown in Table 9.7. We assume that the required transmission rates for a call are different in downlink and uplink, however, their holding time is the same. Note that urban and rural wireless DSL subscriber figures are combined as discussed above.

9.4.2 Subscriber density and offered traffic

We assume that the number of subscribers per square kilometre follows the model developed for the PDC system in Japan: 0.1% of the operator's total subscribers exist in 1 km^2. This subscriber density is assumed for dense-urban areas, where most subscribers access services.

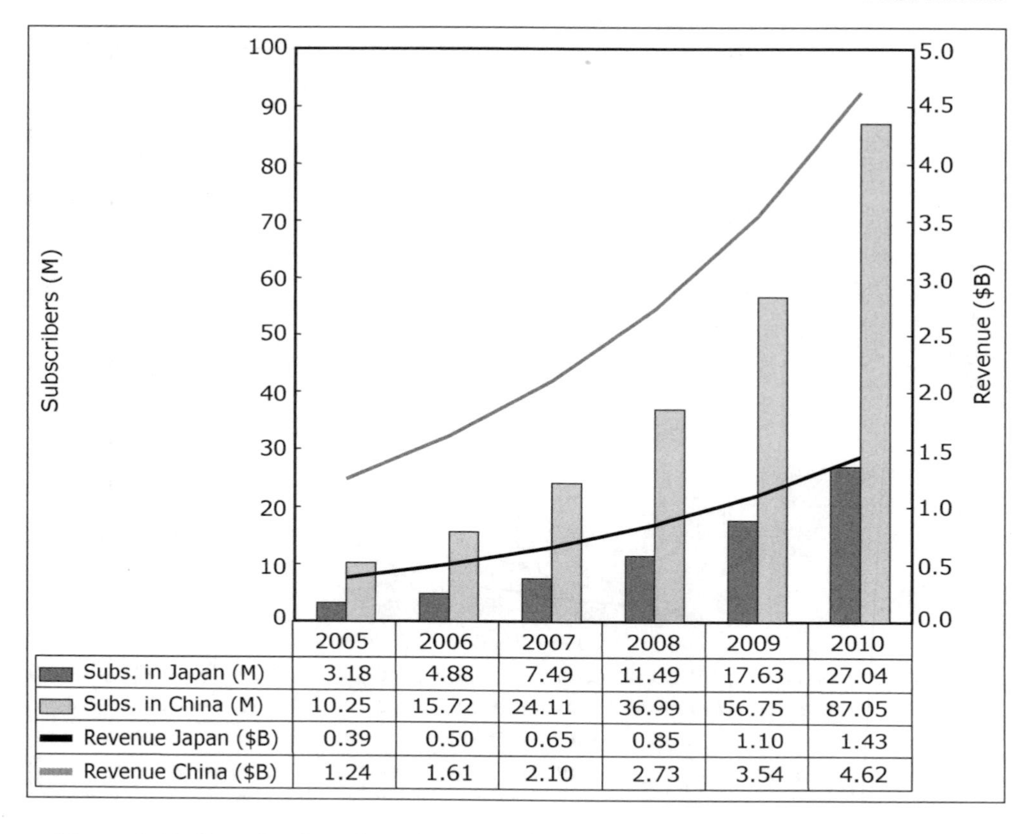

	2005	2006	2007	2008	2009	2010
Subs. in Japan (M)	3.18	4.88	7.49	11.49	17.63	27.04
Subs. in China (M)	10.25	15.72	24.11	36.99	56.75	87.05
Revenue Japan ($B)	0.39	0.50	0.65	0.85	1.10	1.43
Revenue China ($B)	1.24	1.61	2.10	2.73	3.54	4.62

Figure 9.11 Location-based service subscribers and revenue predictions, 2005–2010

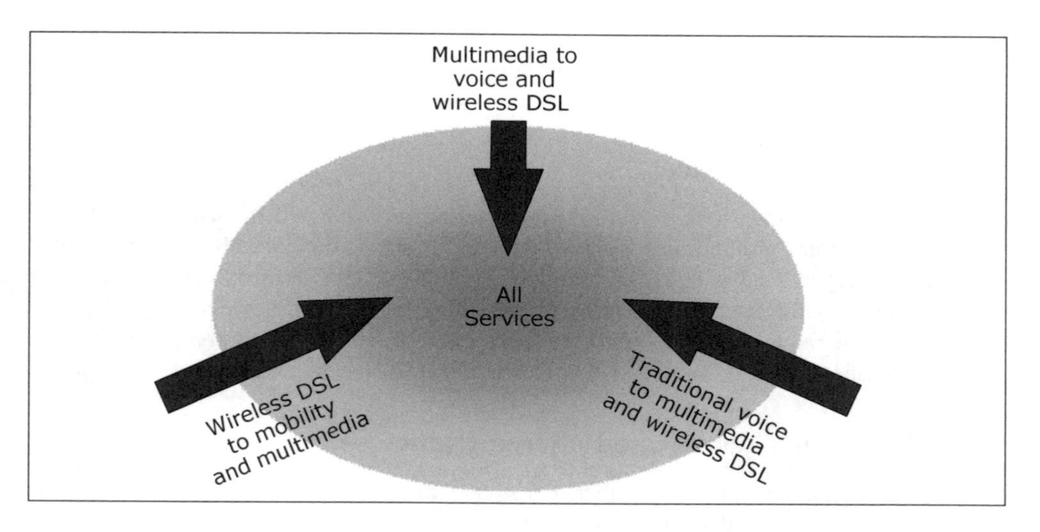

Figure 9.12 Service segments and operator market entry points

Table 9.7 Average transmission rates and call lengths for different services

Services	Transmission rate (kbps) Downlink	Uplink	Average holding time *per peak hour* (s)
Simple voice	16	16	180
Rich voice	64	32	180
Wireless DSL	512	64	240
Customised infotainment	128	16	120
Mobile intranet/extranet	512	128	300
MMS - consumer	16	16	300
MMS - machine-to-machine	8	8	10
Location-based services	64	8	120

For example, an operator with one million subscribers would have a subscriber density of 1000 per km^2 in dense-urban areas. The total offered traffic can then be calculated as based on the average offered traffic per subscriber.

This value can be used to calculate offered traffic figures. For example, the area for one sector in a three-sector cell, with a radius of 400 meters is 0.168 km^2. Such a sector will have users equal to $0.168 \times 0.001 \times$ Total subscribers. Assuming each user generates 0.05 Erlang of traffic (i.e. three minutes of traffic during peak hour), for a service with one million total subscribers, the offered traffic is about 8 Erlangs. That is for a typical cell radius of 400 m, the offered traffic is 8 Erlang per one million subscribers per sector.

9.4.3 Subscriber numbers

Our fictional new operators must compete in a market with several incumbent operators. We assume that a new operator can capture one-third of all new subscribers in each segment where they compete for established services such as voice and MMS. For new services, such as mobile internet, the subscriber estimates are a function of operator strategy and system capacity. The subscriber estimates for each service are shown in respective sections in the following text.

9.5 Costs

We divided the costs associated with a wireless communications service into three sections: (1) costs associated with spectrum acquisition, (2) costs associated with network nodes and devices, and (3) costs associated with operating the network. These were discussed in Chapters 5, 6, and 7. In this section, we present assumptions and assign costs for our fictional operators who may use one of the four technologies stated earlier to provide wireless communications services.

9.5.1 Cost of spectrum

It is very difficult to put a figure on the cost of spectrum. This cost is very much a function of a nation's policy, as well as the number of competing operators and, of course, market characteristics. In countries where market forces decide the spectrum price, through an auction process, highly variable figures have resulted. (An example of the cost of spectrum is the auction figures for 3G spectrum in Europe, which was provided in the box on page 98.) The cost of spectrum in both Japan and China is likely to be very low, as these countries are not expected to auction spectrum for 3G systems and beyond. However, to distinguish between operators using licensed and unlicensed bands we assume spectrum to be licensed by the government at a typical price.

For Japan, we assume a per capita price of $10 for each 10-MHz band. This is somewhat less than what was paid in France for 3G spectrum. For China, we assume a per capita price of $7 per 10-MHz band, although this is calculated only for the urban population. The resulting spectrum costs are shown in Table 9.8. This cost is applicable to each operator using a licensed band with WCDMA, TD-CDMA, and WiMAX technologies. As a WLAN operator uses an unlicensed band, its spectrum cost is zero.

9.5.2 Cost of equipment

As discussed in Chapter 6 the cost of devices is highly variable and is very much a function of volume. We assume the following network nodes:

1. Base stations
2. Network controllers
3. Gateway equipment
4. Access points
5. End-user equipment.

Among these, items 1, 2, and 3 are needed by WCDMA, TD-CDMA, and WiMAX technologies (assuming that WiMAX systems will provide full mobility). The cost of end-user equipment is a function of the device capabilities. WLAN systems are assumed to connect directly to the internet, and therefore have no requirements for items 2 and 3 on the list.

9.5.2.1 Base station costs

In Section 6.1, we defined the different elements of a base station. In our simulations here, we assume that the cost of the baseband module, transport module and control, clock, and power supply modules are the same among all three technologies: WCDMA, TD-CDMA, and WiMAX.

Table 9.8 Cost of 10 MHz spectrum for operators in Japan and China

	Japan	China
Population (M)	128	579 (urban)
Cost per 10 MHz spectrum per capita	$10	$7
Total cost of spectrum ($B)	1.28	4.05

The cost of the RF module depends on the adaptive array antenna requirements and the size of the power amplifier. The overall cost also depends on the IPR cost associated with each technology. Our assumptions on the total cost of a base station are listed as follows:

- WCDMA has the highest IPR cost, requiring a mark-up of around 10%.
- TD-CDMA and WiMAX have a lower IPR cost, resulting in a mark-up of around 5%.
- Adaptive array antenna utilisation requires RF processing, which in turn increases the respective module cost. Base stations with adaptive array antennas are assumed to cost an extra 30%.
- MIMO antenna adaptation requires baseband processing and increases the cost of the respective module cost. Again base stations with adaptive array antennas are assumed to cost an extra 30%.
- TD-CDMA requires only one set of filters for transmission and reception, resulting in slightly decreased RF module cost. In addition, use of smaller power amplifiers leads to a lower base station cost of 10%.
- Although any of the antennas can be used with any of the technologies in our following simulations, only the WiMAX operator is assumed to use adaptive array antennas.

Our cost assumptions for the base stations, including antennas for each of these technologies, are summarised in Table 9.9. We note again that base station cost figures are very much a factor of volume, closely held secrets by both sellers (manufacturers) and buyers (operators), and that these figures are only assumptions.

Further assumed is that the costs of cell design and installation are a function of the number of cell sites, as well as the size of the base station. The latter in particular, depends upon the size of the antennas, the power amplifiers, and the baseband receiver. There is a slight size advantage for using a joint detector and FFT receiver as compared with using rake receiver based systems. The cell design and installation cost assumptions for base stations are listed in Table 9.10.

9.5.2.2 Network controllers and gateway equipment

The cost of these and other nodes, such as the home location register, are again not readily known. We assume that these figures are similar for WCDMA, TD-CDMA, and WiMAX

Table 9.9 Cost assumptions for WCDMA, TD-CDMA, and WiMAX base stations

Cost	WCDMA ($)	TD-CDMA ($)	WiMAX ($)
Equipment modules	50 000	50 000	50 000
IPR	5000	2500	2500
Sector antennas	10 000	10 000	10 000
AAA/MIMO antennas	15 000	15 000	15 000

Table 9.10 Cell design and Installation cost for WCDMA, TD-CDMA, and WiMAX base stations

	Japan ($)	China ($)
WCDMA, TD-CDMA and WiMAX	25 000	20 000

Table 9.11 Device and installation cost assumptions for WLAN in Japan and China

	Japan ($)	China ($)
WLAN access point (device and installation)	5000	4000

technologies. We further assume that the cost is nominally 20% of the total network equipment costs. WiMAX and WLAN systems are assumed to connect directly to the internet at the initial stage, when serving the wireless DSL market. When WiMAX services are extended to provide multimedia and voice services, then these network nodes become necessary additions.

9.5.2.3 Cost of WLAN access points

Access point cost figures are more easily available. We assume a WLAN system with routing capabilities, installed indoors and outdoors. We assume the use of simple antennas with diversity reception capabilities. Total cost assumptions for one access point, including installation, are shown in Table 9.11.

9.5.2.4 End-user equipment costs

The cost of end-user equipment depends greatly on the functionalities of the devices. As shown in Figure 8.1, these devices can take many forms. Our cost assumptions are listed in Table 9.12. A high-end mobile is assumed to include functions such as video and audio streaming, and web browsing. A low-end mobile has only telephony functions. Wireless PCMCIA cards provide no extra functionality of their own and act purely as bridges to the internet. Machine-to-machine MMS units similarly provide only data communications capabilities. Embedded devices, such as WLAN, WCDMA, and WiMAX receivers in

Table 9.12 Cost assumptions for end-user devices in Japan and China

Device type	Japan ($)	China ($)
Multimedia terminal	500	300
Voice terminal	50	50
MMS device	100	100
PCMCIA	50	50

PCs and PDAs also act as wireless bridges. Their cost is assumed to be similar to wireless PCMCIA cards. We assume the end-user equipment costs are similar for both Japan and China.

We note that these costs are either directly or indirectly borne by end-users and not the operator. Subsidies for mobiles or data cards are assumed to exist. These, however, are dealt with in the 'Operating costs' section.

9.5.3 Operating costs

Operating costs were discussed in Chapter 7. These costs are variable, depend on the number of base stations and customers, and recur on a regular (yearly) basis. Among the costs an operator incurs are the following:

1. Maintenance and upgrade costs per base station
2. Cell site rentals
3. Power consumption per base station
4. Fixed-line (optical fibre) leases
5. Customer acquisition costs
6. Subsidy costs per item of user equipment (handset)
7. Billing and customer maintenance
8. Government fees per user
9. Human resources.

We consider items 2, 4, 5, and 6 in our simulations.

9.5.3.1 Cell site rentals

Different types of cell sites are required in urban, suburban, and rural areas. High rooftops are available in many urban areas and may be used for locating base station antennas. The rental for these places vary widely, and follows the general real estate market. In suburban and rural areas, an operator may need to build towers, or rent space on existing towers, to locate the base stations and antennas. Our site rental cost assumptions for base stations and access points are listed in Table 9.13.

Table 9.13 Site rental costs per year in Japan and China

	Japan ($)	China ($)
Base station	15 000	10 000
WLAN access point	1000	1000

9.5.3.2 Fixed-line leasing

Base stations and access points need to be connected to network controllers or to the internet using fixed links. These are generally leased from fixed-line operators and their costs are well known. In rural areas where optical fibres have not been laid, an operator needs to use wireless point-to-point links. Our assumptions on the cost of these wired and wireless links are shown in Table 9.14. We assume that all base stations use a 50:50 mix of fixed/wireless connections. WLANs use leased lines all the time.

9.5.3.3 Costs of customer acquisitions

Customer acquisition costs include advertising and initial complimentary service period fees. Customer acquisition costs are lower in a nonsaturated market segment, whereas in a saturated market they can be quite high. Our assumptions for customer acquisition costs for different market segments in Japan and China are shown in Table 9.15.

9.5.3.4 End-user subsidies

End-user subsidy is the cost an operator pays up front for an end-user's device. These costs can be substantial, with mobile phone subsidies in Japan typically at 400 $\sim$ 500. The subsidy cost differs for different services. Our assumptions for Japanese and Chinese markets are shown in Table 9.16.

Table 9.14 Fixed-line lease per year, and wireless point-to-point costs in Japan and China

	Japan ($)	China ($)
Fixed-line lease (per year)	12 000	10 000
Wireless point-to-point (one-off)	25 000	20 000

Table 9.15 Customer acquisition costs in Japan and China

	Japan ($)	China ($)
Multimedia phone	300	300
Wireless internet/DSL user	400	200
MMS (machine-to-machine)	300	300

Table 9.16 End-user device subsidy assumptions for Japanese and Chinese markets

	Japan ($)	China ($)
Multimedia user	400	100
Wireless internet/DSL user	50	50
MMS (machine-to-machine)	100	100

9.6 Cash Flow Scenarios

In this section, we develop operating business scenarios for Japan and China. As premised, four operators will use four different technologies – WCDMA, TD-CDMA, WiMAX, and WLAN – to start up new telecommunications service businesses. In developing these scenarios, we consider the fact that there are current operators in both markets, and the new operators will need to compete with them for market share in old and new services. We provide two examples in each case. For Japan, we evaluate TD-CDMA and WCDMA, and investigate the advantages of the new technologies. In China we evaluate WiMAX and WLAN, and again investigate the advantages of the new technologies.

9.6.1 Scenarios in Japan

Several established players exist in Japan. We assume our two new operators target the Japanese telecom market as shown in Figure 9.13. The WCDMA operator enters the market to provide voice service at first, but soon expands to provide wireless data services. The TD-CDMA operator focuses on wireless data service, and aims at entering DSL and voice markets in later steps.

9.6.1.1 Operator A: WCDMA

The WCDMA operator is assumed to offer voice and multimedia services, and later on move into the wireless DSL market. The expected subscriber numbers using our earlier assumptions in the eight service market segments are shown in Table 9.17.

The WCDMA operator is assumed to require 20 000 base stations for a full coverage of the populated areas in Japan. The base stations are deployed over a three-year period, with 4000 in the first year, 7000 in the second year, and 9000 in the third year. In busy areas, each base station is assumed to support three sectors.

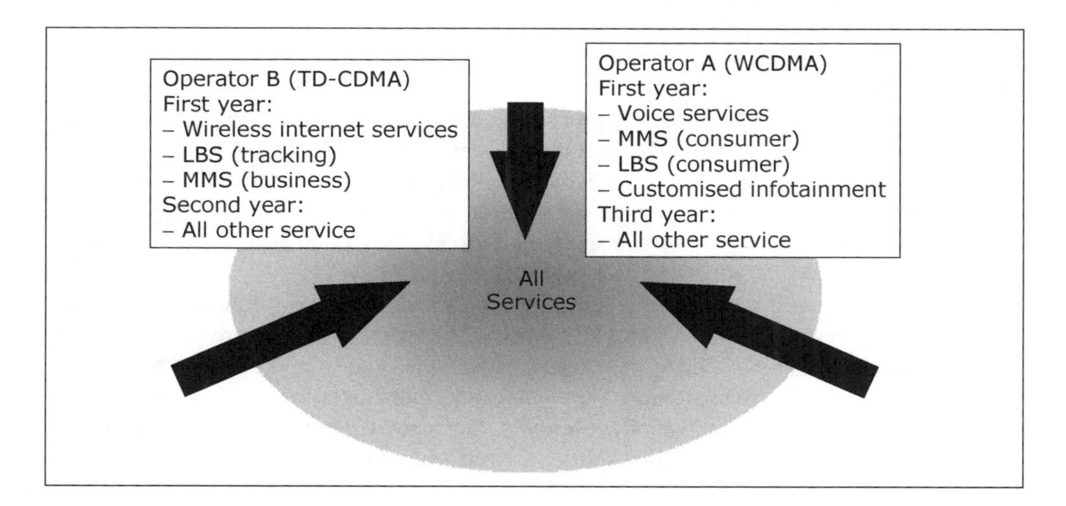

Figure 9.13 Assumptions for two new operators in the Japanese market

Table 9.17 Japanese WCDMA 'Operator A' subscriber numbers (in millions) for the different service segments

Market segment	2005	2006	2007	2008	2009	2010
Simple voice	1.22	2.17	3.15	4.16	4.85	5.56
Rich voice	0.15	0.27	0.39	0.52	0.61	0.70
Wireless DSL	0.00	0.00	0.25	0.88	1.75	2.77
Customised infotainment	0.41	0.72	1.05	1.39	1.62	1.85
Mobile intranet/extranet	0.00	0.00	0.67	1.73	3.41	6.07
Multimedia messaging service, consumer	0.98	1.74	2.52	3.33	3.88	4.45
Multimedia messaging service, machine-to-machine	0.00	0.00	0.50	1.40	3.06	6.06
Location-based services	0.23	0.47	0.77	1.18	1.69	2.44

Table 9.18 Downlink and uplink capacity (in Mbps) for WCDMA for a 10 MHz system

Downlink	Uplink
1.38	0.85

Using a 10-MHz bandwidth (5 MHz down and 5 MHz up), the average throughput from Table 5.5 is found to be 1.38 Mbps in the downlink and 800 kbps in the uplink. Using approximations from Section 2.8.2, the maximum number of simultaneous users that can be supported for simple voice and mobile intranet/extranet for a required grade of service (1% blocking, and 10% queuing probabilities) is calculated. We use required transmission rates figures for uplink and downlink from Table 8.1. These figures, can translate into maximum possible offered traffic in Erlang is found to be 1.38 Mbps in the downlink and 800 kbps in the uplink (Table 9.18).

As shown above, a simple calculation can yield the amount of offered traffic in Erlang in a typical cell in dense-urban areas. If a typical, for Japan, 0.00016% of total subscribers to a service are present in a sector of a cell with 400 meter radius, and each generate 0.05 Erlang of traffic during peak hour (three minutes in one hour), then the total offered traffic is the total number of subscribers of a service multiplied by 0.000008 or eight Erlangs per one million.

The resulting maximum possible users in one cell are given in Table 9.19 for each service. The figures can be used to verify whether the market share figures of Table 9.17 can be supported. Our WCDMA operator, for example, can support voice, MMS, and LBS services with low transmission rates but cannot support wireless DSL and mobile services for the estimated number of subscribers. Our operator will need to use enhancing technologies to support the estimated traffic, as well as extra spectrum.

The cash flow diagram is shown in Figure 9.14.

Table 9.19 Usage traffic capability for each of the service segments for 'Operator A'

Market segment	Required Tx rate (kbps)		Number of channels		Traffic (Erlangs)	
	Down	Up	Down	Up	Down	Up
Simple voice	16	16	86	50	70.2	37.5
Rich Voice	64	32	21	25	12.7	15.9
Wireless DSL	512	64	2	12	0.7	8.4
Customised infotainment	64	16	20	50	15.4	42.6
Mobile intranet/extranet	512	128	2	6	0.7	3.5
Multimedia messaging service, consumer	8	8	172	100	158.2	89.5
Multimedia messaging service, machine-to-machine	4	4	345	200	325.4	185.1
Location-based services	8	2	172	400	158.2	378.9

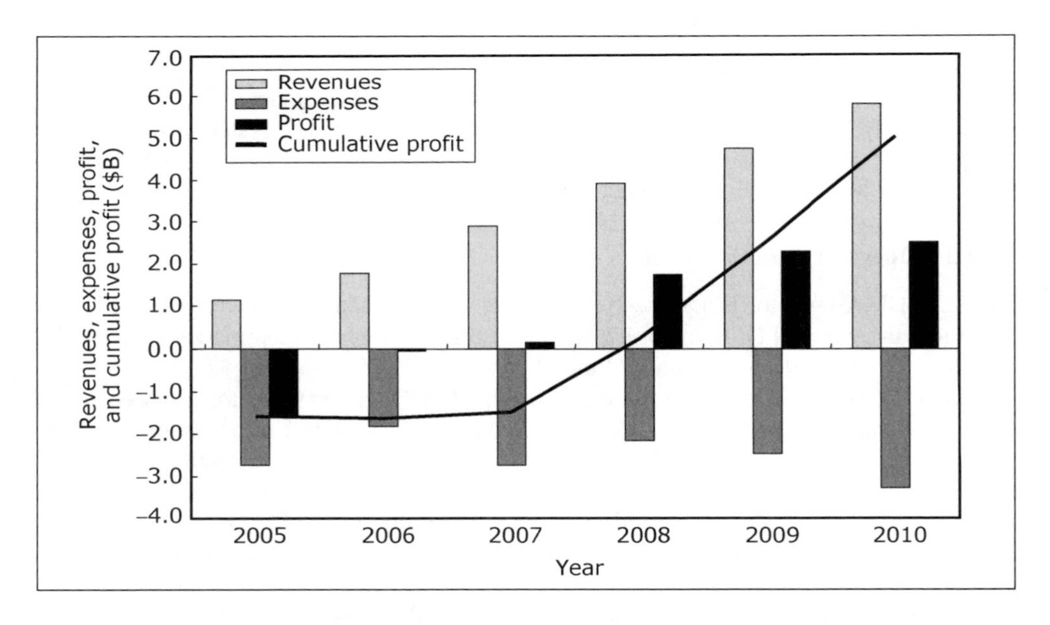

Figure 9.14 Cash flow diagram for 'Operator A'

Technology enhancements

The WCDMA operator can improve its cash flow by using the enhancing technologies discussed in Chapter 3. These improvements can be realised through variations on the following design options. The cash flow diagram is modified accordingly with assumptions from these enhancements.

1. Adaptive array antennas

    - The cost of a base station with these antennas can be higher by about 10%. However, the increased utilisation can result in decreasing the overall number of base stations by an even greater number (e.g. 20%). These figures are dependent, however, on network layout, and the savings on capital expenditure can be even significantly higher.
    - Alternatively, the traffic capacity of individual base stations can be increased. This can result in better service offerings, for example, higher data rates can be offered to data service customers, resulting in a higher market share and higher revenues.
    - A combination of the above.

2. MIMO antennas
   MIMO antennas can be used to increase system capacity or to enhance coverage. Better service quality should lead to higher market share and revenues. Again, figures from Chapter 3 can be used to calculate cost/benefit advantages.

3. OFDM
   The use of OFDM is being considered within the WCDMA standard. This technology can help reduce the level of interference and thereby increase system capacity. This reduced interference and increased capacity leads to higher data rates, with benefits similar to those realised through adaptive array antennas.

9.6.1.2 Operator B: TD-CDMA

The TD-CDMA operator is assumed to offer data communications services, followed by a move into the wireless DSL and voice service segments. The assumed market share in the eight service segments are shown in Table 9.20.

The TD-CDMA operator is assumed to require 18 000 base stations for a full coverage of the populated areas in Japan. The 10% reduction in required base station numbers compared with WCDMA comes from the fact that TD-CDMA's joint detection technology enhances coverage and therefore requires somewhat fewer base stations. The base stations are deployed over a three-year period, with 3000 in the first year, 6000 in the second year, and 9000 in the third year. In busy areas, each base station is assumed to support three sectors.

A 10-MHz bandwidth (TDD, up and down) means that system capacity can be flexibly used between uplink and downlink. The TD-CDMA standard specifies 15 slots in each 10 ms frame as shown in Figure 1.9. Three of these slots are used for control signalling. The remaining 12 slots are assigned to down- or uplink traffic. The DL:UL ratio can be set anywhere between 11:1 and 1:11. The TD-CDMA system capacity for several allocations is shown in Table 9.21. Again, the maximum possible number of users can be calculated for each service by calculating the trunking efficiency as discussed in Section 2.8.2. Following the same calculation as for 'Operator A', our TD-CDMA operator also can support voice and other low speed services, and to a larger degree as compared with Operator A's Wireless DSL services. Nevertheless, 'Operator B' also will need enhancing technologies to support the estimated traffic. The results are shown in Table 9.22.

Table 9.20 Japanese TD-CDMA 'Operator B' subscriber numbers (in millions) for the different service segments

Market segment	2005	2006	2007	2008	2009	2010
Simple voice	0.00	0.95	1.93	2.94	3.63	4.34
Rich voice	0.00	0.12	0.24	0.37	0.45	0.54
Wireless DSL	0.01	0.06	0.31	0.94	1.81	2.83
Customised infotainment	0.00	0.32	0.64	0.98	1.21	1.45
Mobile intranet/extranet	0.27	0.69	1.35	2.41	4.09	6.76
Multimedia messaging service, consumer	0.00	0.76	1.54	2.35	2.91	3.47
Multimedia messaging service, machine-to-machine	0.15	0.42	0.92	1.83	3.48	6.48
Location-based services	0.23	0.71	1.32	2.13	3.16	4.65

Table 9.21 Downlink and uplink capacity (in Mbps) for TD-CDMA with several slot allocation ratios for a 10-MHz system

Slots allocated		Capacity	
Downlink	Uplink	Downlink	Uplink
11	1	5.47	0.37
9	3	4.10	1.10
6	6	2.73	2.20
3	9	1.37	3.30
1	11	0.46	4.03

Table 9.22 Usage traffic capability for each of the service segments for 'Operator B', with 9:3 slot ratio

Market segment	Required Tx rate (kbps)		Number of channels		Traffic (Erlangs)	
	Down	Up	Down	Up	Down	Up
Simple voice	16	16	256	68	231.8	53.7
Rich Voice	64	32	64	34	50.0	23.5
Wireless DSL	512	64	8	64	5.1	55.6
Customised infotainment	64	16	64	68	55.6	59.3
Mobile intranet/extranet	512	128	8	8	5.1	5.1
Multimedia messaging service, consumer	8	8	512	137	488.1	124.7
Multimedia messaging service, machine-to-machine	4	4	1024	274	990.1	256.6
Location-based services	8	2	512	550	488.1	525.2

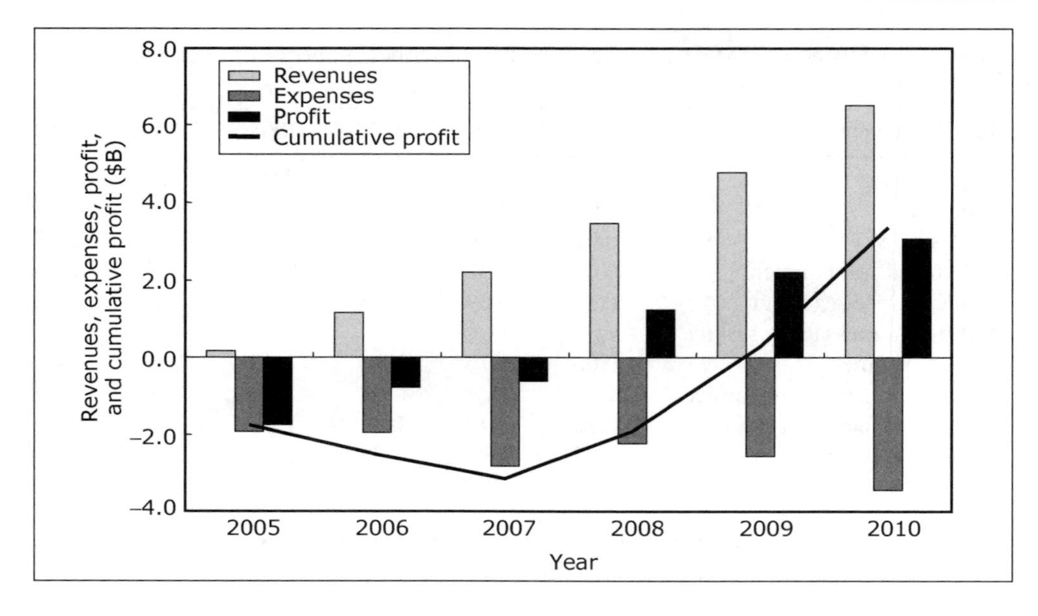

Figure 9.15 Cash flow diagram for 'Operator B'

The cash flow diagram is shown in Figure 9.15.

Technology enhancements

The TD-CDMA operator can also improve its cash flow by using the enhancing technologies discussed in Chapter 3. Here again, benefits are gained from the use of adaptive array antennas, or MIMO antennas. In contrast to WCDMA, the introduction of OFDM does not bring significant benefits to TD-CDMA. The reason is that joint detection technology already reduces much of the intra-cell interference, and OFDM introduction cannot enhance this further without using more expensive power amplifiers.

One possible enhancement is to use the mini base station concept discussed in Section 4.4. With an overlay design structure, the infrastructure cost can be significantly reduced. This in turn positively improves the cash flow profile.

9.6.2 Scenarios in China

Much like the examples for Japanese operators, similar scenarios can be developed for China. In China, the scenarios concentrate on WiMAX and WLAN technologies. The two new operators here target the Chinese telecom market as shown in Figure 9.16. Our WiMAX operator aims at the data services market, wireless DSL for rural and urban subscribers, and later moves into the multimedia and voice service markets. The WLAN operator targets only the data market with wireless DSL and intranet/extranet internet services.

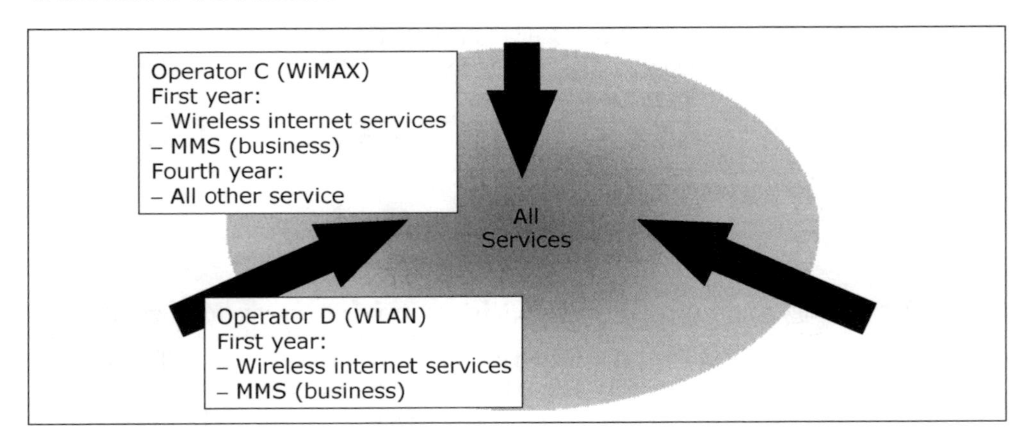

Figure 9.16 Assumptions for two new operators in the Chinese market

9.6.2.1 Operator C: WiMAX

The WiMAX operator is assumed to initially offer wireless DSL and intranet/extranet internet data services, and then move into multimedia data communications and the voice service segments. The assumed market share in the eight service segments are shown in Table 9.23.

The WiMAX operator is assumed to require 60 000 base stations for a full coverage of the populated areas in China. The base stations are deployed over a three-year period, with 15 000 in the first year, 20 000 in the second year, and 25 000 in the third year.

Assuming an FDD operation using a 10-MHz bandwidth (5 MHz down and 5 MHz up), the assumed average throughput is shown in Table 9.24. On the basis of these figures, we calculate the maximum number of users, as in Section 2.8.2, that can be supported. These

Table 9.23 Chinese WiMAX 'Operator C' subscriber numbers (in millions) for the different service segments

Market segment	2005	2006	2007	2008	2009	2010
Simple voice	0.00	0.00	0.00	12.68	24.67	34.92
Rich voice	0.00	0.00	0.00	1.59	3.08	4.37
Wireless DSL	0.01	0.33	1.49	4.49	8.09	12.14
Customised infotainment	0.00	0.00	0.00	3.17	6.17	8.73
Mobile intranet/extranet	0.71	1.84	3.63	6.47	10.98	18.13
Multimedia messaging service, consumer	0.00	0.00	0.00	4.44	8.63	12.22
Multimedia messaging service, machine-to-machine	0.26	0.81	1.95	4.33	9.32	19.76
Location-based services	0.00	0.00	0.00	4.34	9.50	15.80

Table 9.24 Downlink and uplink capacity (in Mbps) for WiMAX for a 10 MHz system

Downlink	Uplink
3.0	0.6

Table 9.25 Usage traffic capability for each of the service segments for 'Operator C'

Market segment	Required Tx rate (kbps)		Number of channels		Traffic (Erlangs)	
	Down	Up	Down	Up	Down	Up
Simple voice	16	16	187	37	165.5	26.1
Rich Voice	64	32	46	18	33.9	10.3
Wireless DSL	512	64	5	9	2.7	5.9
Customised infotainment	64	16	46	37	38.9	30.7
Mobile intranet/extranet	512	128	5	4	2.7	2.1
Multimedia messaging service, consumer	8	8	375	75	354.6	65.9
Multimedia messaging service, machine-to-machine	4	4	750	150	721.0	137.1
Location-based services	8	2	375	300	354.6	281.8

user numbers are listed in Table 9.25 for each service. Similarly, this data, in conjunction with subscriber distribution figures for China, can be used to verify whether the market share figures of Table 9.23 can be supported.

The cash flow diagram is similarly drawn for our WiMAX operator in Figure 9.17.

Technology enhancements

The WiMAX operator can similarly improve its cash flow by using the enhancing technologies discussed in Chapter 3. Here again, benefits are realised from the utilisation of adaptive array antennas or MIMO antennas. As WiMAX already uses OFDM, no further enhancement will be possible.

9.6.2.2 Operator D: WLAN

The WLAN operator is assumed to offer wireless DSL services in rural areas and intranet/extranet internet data services. It is assumed that Operator D does not provide full coverage and therefore will not compete in other service segments. Other assumptions, similar to this (here 200 000 access points or more may be reasonable), and assumptions on market share from data services, can be used to calculate cash flows.

The cash flow diagram for 'Operator D' is shown in Figure 9.18.

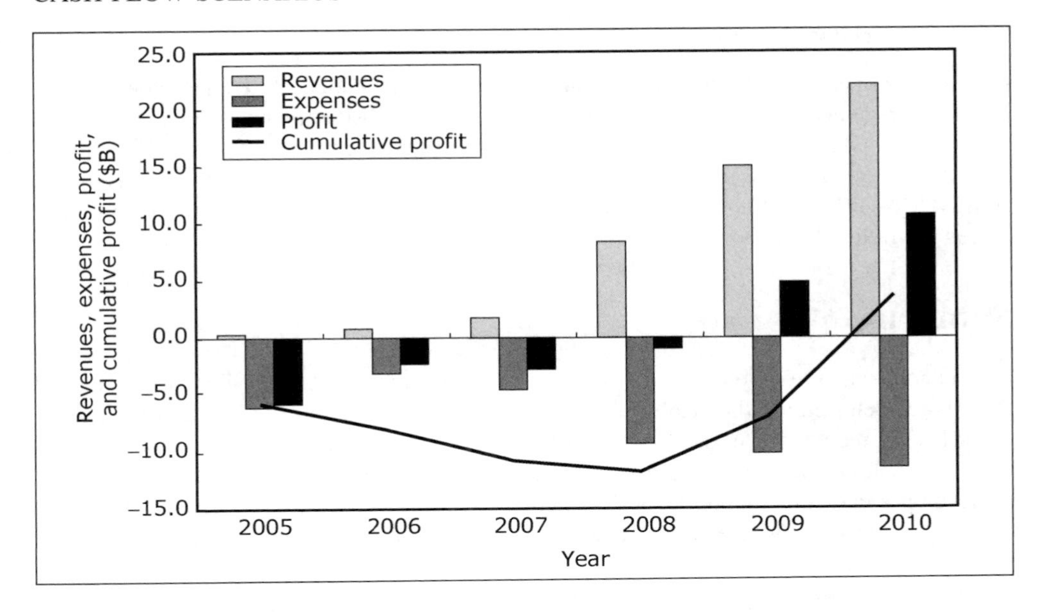

Figure 9.17 Cash flow diagram for 'Operator C'

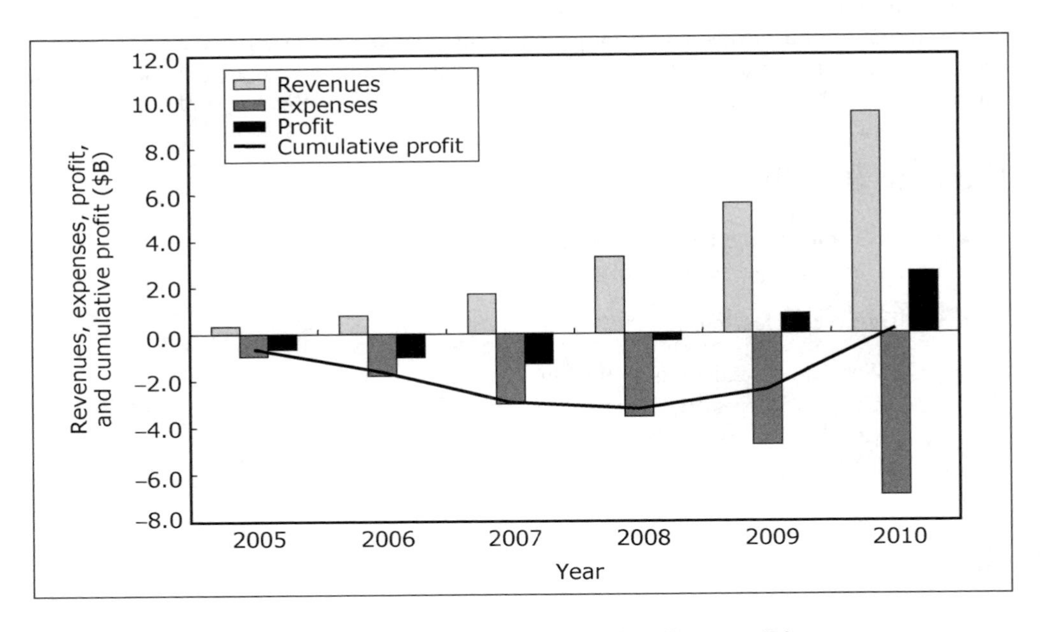

Figure 9.18 Cash flow diagram for 'Operator D'

Technology enhancements

Our WLAN operator can improve its cash flow by using the enhancing technologies. Again, benefits are gained by using adaptive array antennas or MIMO antennas. Because WLAN already uses OFDM, no further enhancement is possible here.

The cost of backbone fixed- link lease is a major part of a WLAN operating expenditure. A WiMAX–WLAN networking topology can be used to reduce these costs. In fact, in many places a wireless backbone may be the only option.

Summary

This chapter has presented a model for developing business scenarios. First we discussed models for determining the size of a target market. This modelling, in addition to creating a market segmentation, is used to estimate market size, possible revenues, and required network deployment. We then analysed the costs involved and assumed certain cost figures to develop a cash flow diagram for four fictitious operators, two each in Japan and China. We then listed areas where technology can be used to improve network operation and to improve cash flow.

Although the figures presented here are typical, the scenarios developed are not intended to depict reality. Thankfully, all of our operators have ended-up with positive cash flow. It may not be so in real life. Our purpose here has been to familiarise the reader with the steps that go through a business plan development. A number of cost and income sources have been omitted, interest rates have been assumed to be zero, and so on. The models, indeed, can and need to have more details. These are, however, outside the scope of this book.

Further Reading

- United Nations Population Division web site:
  `http://esa.un.org/unpp/`, 2005
- On Traffic engineering:
  - Web site for calculating blocking and queuing probabilities:
    `http://personal.telefonica.terra.es/web/vr/erlang/eng/cerlangc.htm`, 2005

Glossary

1G	The First-Generation, analogue mobile communications systems, including AMPS and NMT.
16-QAM	16 Quadrature Amplitude Modulation, a technique for converting data bits to send over transmission channels.
2G	The Second-Generation, digital mobile communications systems, including GSM and IS-95.
3G	The Third-Generation, digital mobile communications systems, including WCDMA and TD-CDMA.
3GPP	The Third-Generation Partnership Project, a body standardising 3G systems.
4G	The Fourth Generation of mobile communications systems, systems that are expected to appear after 3G.
AAA	Adaptive array antennas.
Access point	A wireless LAN node that routes data streams from WLAN cards to the internet and vice versa.
Access technologies	Technologies used for accessing the network: CDMA, TDMA, and so on.
Access-focused	A wireless service category for services that connect an end-user to the internet.
ACM	Adaptive coding and modulation, a technique used to increase throughput of the communications channels.
ADSL	Asynchronous Digital Subscriber Link, a technology for high-speed data transmission using telephone wires.
ADSL-2	An enhanced ADSL standard that increases the transmission rates through more efficient coding and modulation techniques.
ADSL-2+	An enhancement to ADSL-2 standard that specifies a doubling of the downstream transmission rates.
Air time	The time for which a user occupies the channel.
AMPS	Advanced Mobile Phone Service of North America; a 1G analogue system.

Broadband Wireless Communications Business: An Introduction to the Costs and Benefits of New Technologies Riaz Esmailzadeh

AMR	Adaptive Multi-rate; a voice encoder technology that has been standardised for 3G services. AMR has eight different compression ratios resulting in compressed bit rates of 4.75 kbps to 12.2 kbps. There is also an enhanced AMR.
AP	A wireless LAN 'Access Point'.
ARPU	Average Revenue per User, a measure of operator revenues.
ARQ	Automatic Repeat reQuest, a technique that ensures data packets are communicated correctly to a receiver.
ASIC	Application-specific Integrated Circuits.
ATM	Asynchronous Transfer Mode, a mode of transporting data packets.
Attenuation	A decrease in signal level.
Back-haul operator	A company that owns/operates fixed links (usually optical fibre), and leases them to other telecom operators.
Bandpass filter	An electronic circuit that passes signals within a certain frequency band, and stops all out-of-band signals.
BER	Bit Error Rate, a measure of communication quality.
Bit/sec/Hz	Frequency-utilisation efficiency, expressed in the number of information bits that can be sent in 1 s, normalised over 1 Hz of bandwidth.
BLER	Block Error Rate, a measure of communications quality.
BPSK	Binary Phased Shift Keying, a technique for converting data bits to send over transmission channels.
Broadband wireless	Refers to high transmission rate services over wireless systems; compares with fixed broadband services carried over wired systems.
BREW	A platform software created by Qualcomm.
Browser	A software tool that facilitates viewing world-wide web pages.
BS	Base Station.
CAGR	Compound Annual Growth Rate, a measure that indicates average growth of a parameter over several years.
CDMA	Code Division Multiple Access, a technology for 2G and 3G systems.
CDMA-2000	Code Division Multiple Access-2000, a 3GPP2 standard.
Channel coherence bandwidth	The bandwidth over which the fading behaviours of all frequency components are highly correlated.
Channel reciprocity	A condition wherein channel characteristics between the uplink and downlink are reciprocal; occurs in TDD-based systems.
Chase combining	In the Chase combining technique, an erroneously received packet is combined with its retransmitted replica.
Chip rate	The number of chips per second for a CDMA system.

Chips/sec	See Chip rate.
Circuit switched	A mode of resource allocation wherein an arriving call is assigned a circuit for use in the entire duration of the call; used in telephony services.
Clipped	The nonlinearity and signal distortion leading to reception error that has resulted from an input signal being larger than the maximum input power level.
Co-channel interference	Interference on a desired user signal deriving from all other users' signals that occupy the same frequency band.
Code domain equalisation	Equalising a signal to remove the effect of co-channel interference due to loss of code orthogonality and channel multi-path effects.
Coding rate	In forward error correcting coding systems: the ratio of information bits to the entire number of bits transmitted.
Consumer navigation	A location-based service in which the network operator helps end-users find a physical location.
Coverage	The total area a mobile network covers, that is, provides effective wireless services.
CPE	Customer Premise Equipment, end-user equipment/devices.
CSMA	Channel Sense Multiple Access, a technology used in wireless LAN systems.
D-AMPS	Digital American Mobile Phone System, a 2G digital system.
Data sink	Data destination; opposite of data source.
DECT	Digital Enhanced Cordless Technologies, a 2G European portable/mobile communications systems.
Delay spread	The time difference between arrival of the first significant ray and the last significant ray of the same signal transmitted over a wireless channel. Signals arrive at different times as they travel different routes and get reflected from natural objects.
Diversity reception	Receiving signals from diverse routes, for example, from two separate antennas.
DL	Downlink (also called forward link).
DSL	Digital Subscriber Link, a technology used to deliver high-speed data to an end-user.
Duplexing	Sending communications in opposite directions at virtually the same time.
E-commerce	Electronic commerce; business done over the internet.
Edholm's law	Predicts a convergence in the transmission rates of fixed-wire and wireless traffic.
Equalisation	A process to compensate for a channel's time and frequency variations.

Erlang-B formula	A mathematical equation calculating blocking probability from offered traffic and a number of circuits.
Erlang-C formula	A mathematical equation calculating queuing probability from offered traffic and a number of circuits.
ETSI	European Telecommunications Standards Institute, a body for standardising information and communications technologies.
FDD	Frequency Division Duplexing; using two carrier frequencies to transmit in opposite directions.
FDMA	Frequency Division Multiple Access; a technology used in 1G analogue systems.
FEC	Forward Error Correction coding.
FER	Frame Error Rate, a measure of communication quality.
FFT	Fast Fourier Transform (see Fourier Transform).
FFT Receiver	A kind of receiver used in OFDM-based systems.
Fixed network	The wired (optical fibre or twisted copper wire) network, as opposed to wireless.
Flash memory	A type of card that can be inserted into PCs; among other uses, the flash memory card can be used for memory storage or as a wireless bridge.
Fourier transform	A mathematical process for converting a signal presentation from time domain to frequency domain and vice versa.
Frequency domain equalisation	Equalising in the frequency domain by measuring and controlling the level of different frequency components.
Frequency reuse	A cell that uses the same carrier frequencies as used in another cell; such usage is expressed as a ratio.
Frequency-selective fading	A fading process where different frequencies undergo a different, almost independent, signal variation.
FTTH	Fibre-to-the-home services (also FTH).
Gateways	Nodes that interface legacy systems – gateway to the public analogue telephone network; and nodes associated with authentication and with billing.
GGSN	Gateway GPRS Support Node – a gateway where a packet call is switched to the public internet and vice versa.
GPRS	GSM Packet Radio Service.
GSM	Global System Mobile, a 2G mobile communications technology.
Guard time	The time between two OFDM symbols, or between two TD-CDMA time slots.
HARQ	Hybrid Automatic Repeat Request.
HLR	Home Location Register.
Hot Spot service	Providing high data-rate broadband services only in limited high traffic areas, such as train stations, coffee shops, malls, and so on.

Hybrid ARQ	Hybrid Automatic Repeat reQuest, an advanced ARQ technique.
IBO	Input-Back-Off – The practice of backing-off by several dB from the maximum input power level, in a power amplifier, in order to avoid clipping and signal distortion; power amplifiers must not operate at maximum input power levels.
IEEE	The Institute of Electrical and Electronic Engineers, Inc., a professional society, which also standardises telecommunications technologies.
IEEE 802	A standard group within IEEE.
IEEE 802.11	A sub-group within IEEE 802 standardising an Air Interface for wireless LAN Systems.
IEEE 802.16	A sub-group within IEEE 802 standardising an Air Interface for Fixed Broadband Wireless Access Systems.
IEEE 802.20	A sub-group within IEEE 802 standardising an Air Interface for Mobile Broadband Wireless Access
i-Mode	An email/short messaging service in Japan that also provides access to information via an internet browser; very popular; copied in Europe.
IMT-2000	The IMT-2000 Plan Band supports both FDD and TDD.
IMT-extension band	One of several frequency bands allocated for extending IMT services (when the present spectrum is exhausted).
Infotainment	Information and entertainment services combined or offered simultaneously.
IP	internet Protocol.
IP backbone system	A system with a back-haul based on IP.
IP Network	Internet Protocol Network.
IP Router	Internet Protocol Router.
IPR	Intellectual Property Rights (patents and such issues).
IPWireless	The leading company to develop broadband base stations and mobiles according to 3GPP TD-CDMA standards.
IR	The Incremental Redundancy technique takes advantage of variable FEC coding rates. Initially in the IR technique, a low coding rate is chosen for transmission.
IS-95	Interim Standard No. 95, a North American 2G standard.
ISDN	Integrated Services Digital Network, a technology for the transmission of digital voice over ordinary phone lines.
ISM Band	An unlicensed frequency spectrum allocated for Industrial, Scientific, and Medical purposes.
ISP	Internet Service Provider.
IT	Internet and Telecommunications.
ITU	International Telecommunication Union, a body under the United Nations.
Java	A platform software created by Sun Microsystems.

JD	See Joint Detection.
Joint detection	A technique to detect multiple received users' signals, also multi-user detection (MUD).
LAN	Local Area Network.
LDPC codes	Low-density Parity-Check codes, a powerful FEC coding technique.
Link budget	A technique for calculating the received signal level.
Location-based services	A class of services relying on, and oriented to, the importance of the location of an end-user.
M-commerce	Mobile commerce; similar to e-commerce.
m-PSK	Multi-phase Shift Keying (see BPSK and QPSK).
m-QAM	Multi-quadrature Amplitude Modulation (see 16-QAM).
MAC layer	Media Access Control layer, a part of where resource allocation functionalities are performed.
Macro-cell	A call area with a radius of several hundred meters to a few kilometers, covered by a single base station.
Mailbox units	A service that sends an SMS to a mobile notifying an end-user's account of an email arrival.
Mbps	Mega bits per second.
MBps	Mega bytes per second.
Mcps	Mega chips per second.
MIC	Japanese Ministry of Internal Affairs and Communications.
Micro-cell	A cell area in the order of one-tenth to one-twentieth the umbrella of a macro-cell ($50 \sim 300\,\text{m}$).
MIMO antennas	Multi-input, Multi-output Antennas.
Mini-cell	An underlay cell; it differs from a micro-cell in that it can be operated with the area coverage of a macro-cell, even though the two cell areas are not mutually isolated.
MMDS	Multimedia Data Service, a broadband wireless packet-switched service.
MMS	Multimedia Messaging Service, a data service in which users send short messages with a picture or a video clip attached.
Mobility-centric	A class of services that is useful to an end-user because of the user's mobility.
Mobile	Users who require connection to the network/internet at all times, even while travelling to a destination.
Mobile intranet access	A service category for users who access a corporate network using a wireless device.
Mobile internet access	A service category for users who access the internet using a wireless device.
Mobile extranet access	A service category for users who access an external network using a wireless device.
Modulation	The process for converting a digital bit stream, ones and zeros, to patterns that can be transmitted over a channel.

MOS	Mean Opinion Score, a measure related to PQoS. Its parameter is between one and five with five representing excellent quality and one representing poor quality.
MUD	Multi-user Detection.
Multi-user detection	See Joint detection.
Multi-hop	Communications between a base station and an end-user via another base station or mobile device.
MVNO	Mobile Virtual Network Operator, a wireless operator who does not own a network but rents capacity from another operator.
Network nodes	The devices and gateways that convey a data stream between a base station and the public network/the internet.
NB-AMR	Narrow-band AMR.
NMT	Nordic Mobile Telephony (started in Denmark in 1979).
Nomadic	Users who carry their data terminals and connect to the internet from different places. The users require portability and want connectivity after arriving at a place, and not necessarily while travelling there.
NTT	Nippon Telegraphy and Telephone.
OBSAI	Open Base Station Architecture Initiative, an industry group that tries to 'create an open market for cellular base stations' and reduce the cost of development and manufacturing through specifying base station modules and their interfaces.
OFCDMA	Orthogonal Frequency-Code Division Multiple Access.
OFDM	Orthogonal Frequency Division Multiplexing.
OFDMA	Orthogonal Frequency Division Multiple Access.
O&M	Operations & Maintenance.
P2P	Peer-to-Peer; refers to services and software that allow users to share files over the internet.
Packet-switched	The transmission of a data stream by dividing it into a number of packets and transmitting the packets one by one.
PAPR	Peak Average Power (amplitude) Ratio
PC	Personal Computer.
PCMCIA	Personal Computer Memory Card International Association, a standardised type of card commonly used in PCs.
PDA	Personal Data Assistant, a device that provides many functions including a diary and memory.
PDC	Personal Digital Cellular, a 2G mobile communications systems in Japan.
Peta bit	10^{15} bits.
PHS	Personal Handyphone Systems, a 2G Japanese mobile communications systems.

Physical layer	The level at which data transmissions, in a modulated form, take place.
Pico-cell	Smaller than a micro-cell and generally in an isolated, confined area. This could be in an auditorium, inside a plane, or on a train platform.
Portal	An entry point to the internet; in wireless communications it is usually the home page of the mobile operator.
Portal-focused	A service category that serves the needs of users at their entry access point to the internet.
Power control	The process of adjusting transmission power levels in order to compensate for communications channel variations.
PQoS	Perceptual Quality of Service; measured in units of mean opinion score (MOS); see also MOS.
Push-to-talk	A wireless voice service between multiple end-users where one user talks and everyone else listens.
QoS	Quality of Service.
QPSK	Quadrature Phase Shift Keying, a technique for converting data bits to send over transmission channels.
Rake receiver	A type of receiver used in CDMA systems.
Reception diversity	Receiving replicas of the same signal from independent routes.
RF	Radio Frequency.
RF module	Radio Frequency module.
RF processing	Radio Frequency processing
Rich voice	A voice service category enriched by video, photos, and/or higher voice quality.
RNC	Radio Network Controller, a node that controls several base stations.
SF	Spreading Factor.
SGSN	Serving GPRS Support Node; switches packets from a user in the service area to and from the internet.
Shannon theorem	A theorem that calculates the maximum possible capacity of a communications channel as a function of channel quality and bandwidth.
Signal clipping	Signal distortion caused by processing through a nonlinear power amplifier, wherein different signals are amplified by different degrees.
Signal fading	A fading channel refers to a channel where the received signal level varies from large values to very small (faded) values.
Simple voice	A service category providing voice telephony.
SINR	Signal-to-Interference and Noise Ratio.
SIR	Signal-to-Interference Ratio.
SMS	Short Messaging Services, a service where mobile users send text messages to other users.

SNR	Signal-to-Noise Ratio.
Spreading code length	The length of a CDMA spreading code in numbers of chips.
Symbol	A communications unit that has a fixed length, and carries data bits
Symbol length	The time length of a symbol in seconds.
System capacity	The total number of users a system can serve at any particular time.
TACS	Total Access Communication System, a wireless communication technology introduced in mid-1980s.
TD-CDMA	Time Division Code Division Multiple Access, a 3GPP standard.
TD-SCDMA	Time Division Synchronous Code Division Multiple Access, a 3GPP standard.
TDD	Time Division Duplexing; using different time slots to transmit in opposite directions.
TDMA	Time Division Multiple Access, a technology used mostly in 2G systems.
Throughput	The amount of information that can be transported over a communications channel.
Time domain equalisation	Equalisation in the time domain by means of power control.
Transmission diversity	Transmitting from two antennas, as opposed to reception diversity (see the preceding text).
Transport module	A module in a base station that controls the flow of data to and from fixed links.
Turbo codes	A class of powerful FEC codes.
UE	User Equipment; also end-user equipment.
UL	Uplink (also called Reverse Link).
UMTS	Universal Mobile Telecommunications Services, a terminology used by the ITU.
UMTS 3G	UMTS third-generation standards.
UMTS forum	An industry forum focused on UMTS 3G technologies, standards, and services.
Usage time	The time a user is on air (uses a network resource).
Vocoder	Voice encoder, a device that converts an analogue voice stream into a compressed digital stream.
VoIP	Voice over Internet Protocol, a packet-switched voice technology.
WAP	Wireless Access Protocol, a protocol for viewing web sites on mobile phones.
WB-AMR	Wide-band-AMR.
WCDMA	Wide-band Code Division Multiple Access, a 3GPP FDD standard.
WiBro	A pre-WiMAX system developed in Korea.

Wide-area	Providing high data-rate broadband service over a wide geographic area.
WiFi	Wireless Fidelity; refers to WLAN systems.
WiMAX	World-wide Inter-operability for Microwave Access, a technology for fixed, portable, and mobile communications.
Wireless bridges	Devices that create a wireless link between the end-user device (e.g. a PC) and the internet.
WLAN	Wireless Local Area Network.
WWRF	Wireless World Research Forum; WWRF brings together experts from the industry and academia 'to formulate visions on strategic future research directions in the wireless field. ... Our mission is to shape the global wireless future'.
XHTML	Extensible Hyper-Text Markup Language, a script language similar to HTML, which is used extensively in constructing web pages.

Index

Broadband Wireless Communications Business: An Introduction to the Costs and Benefits of New Technologies Riaz Esmailzadeh